Quantum Philosophy

✜

Quantum Philosophy

UNDERSTANDING AND INTERPRETING

CONTEMPORARY SCIENCE

✛

ROLAND OMNÈS

Translated by Arturo Sangalli

PRINCETON UNIVERSITY PRESS

PRINCETON, NEW JERSEY

Translated from the French edition of Roland Omnès,
Philosophie de la science contemporaine
(Paris: © Editions Gallimard, 1994)

Library of Congress Cataloging-in-Publication Data
Omnès, Roland
[Philosophie de la science contemporaine. English]
Quantum philosophy : understanding and interpreting contemporary
science / Roland Omnès ; translated by Arturo Sangalli.
p.   cm.
Includes bibliographical references and index.
ISBN 0-691-02787-0 (cl. : alk. paper)
1. Physics—Philosophy.   2. Mathematics—Philosophy.
3. Science—Philosophy.   4. Quantum theory.
I. Title.
QC6.O55   1999
530′.01—dc21   98-42445

Published with the assistance of the French Ministry of Culture

This book has been composed in Sabon

The paper used in this publication meets the minimum requirements
of ANSI/NISO Z39.48-1992 (R1997) (*Permanence of Paper*)

http://pup.princeton.edu

Printed in the United States of America

3   5   7   9   10   8   6   4

⁘ *Contents* ⁘

CONTENTS

CONTENTS

✢ *Preface* ✢

THIS ESSAY has a leading thread, whose origins may be retraced to Francis Bacon's *The Great Instauration*: one day, the principles of science will be so close to the heart and the essence of things that philosophy will be able to find in them its own foundations. Let us temper that wish and speak only of philosophy of knowledge; let us, on the contrary, bolster it and say that such a day has arrived, and there you have the summary of this book.

The time has come to force our way out of a current crisis in epistemology. There is indeed a crisis, for unlike the flourishing situation in the history of knowledge, the philosophical reflection about science has lost its way—or stagnates. The fashionable authors see only uncertainties, paradigms without enduring principles, an absence of method, and a presence of erratic revolutions, precisely when we should be trumpeting the success of a science whose extent and consistency are unprecedented. To counter this deficiency we can turn only to ancient thinkers, no doubt wiser, but also unable to provide the required antidote, for their science is no longer ours; it has progressed too much.

Beyond the shadow of a doubt, the origin of this crisis is to be found in an event that no one has fully recognized in all its significance: the irresistible irruption of the formal approach in some fundamental sciences such as logic, mathematics, and physics. As a consequence, these disciplines have become practically impenetrable, which explains the capitulation or the adventurousness of so many commentators, not to mention the disarray of the honest man or woman who wonders what those who should understand these subjects are talking about.

A good part of this book retraces this rise toward formalism and shows its necessity, not only in mathematics, but also in the foundations of relativity and quantum physics, and in the theories dealing with all that makes up the universe, space, and particles. As a counterbalance, another part of the book shows how to loosen that formalism and overcome it. The path was shown by certain advances in the interpretation of quantum mechanics, thanks to which it was possible to resolve a good number of difficulties that were hard to accept even in this domain where, more than in any other, the principles of philosophy clash with those of nature. The key to the problem appeared only gradually, through efforts in specialized fields and technical results. But in the end, everything turned out to be quite simple: the principles that science has already mastered are sufficient to recover common sense, to demonstrate its necessity in a certain sense, and at the same time to establish its limits and those of certain philosphical "principles" derived from it. Thus, despite its formal aspects, science brings with it a theory of knowledge, once again transparent, that can explain how we humans understand the world.

Could all that ever lead to a philosophy of knowledge reaching into the very nature of reality? We do not know, even if we can see it taking form already, while we are still busy only dreaming of it.

✤ *Acknowledgments* ✤

Those persons who have inspired me, helped me, or corrected
me in the course of time are too numerous to be named. Thus I will
thank here only Liliane, my wife, whose patience stretched well
beyond the time required for the writing of these pages: all through
those long years, going back well into the past, that I spent trying
to understand.

Few changes or additions have been made in this English translation. I very much enjoyed the exchange of ideas I had with Arturo
Sangalli, who often made the present text better than its initial
French version. I also wish to say how happy I am to see this book
published by Princeton University Press, which certainly stands as
a landmark in the world of learning.

⁓ *Prelude* ⁓

WE WERE IN HELL, or rather in Hades. It is a pleasant place, and I had entered it by chance. Cerberus' question, "Who are you?," had first baffled me, but I was lucky to have answered, "A son of Pan's." The logic was impeccable: Pan conceived the fauns, who in turn did what it takes to conceive, not only among themselves, and abundantly so. Then, a simple calculation of the odds would confirm that I was descended from them. Cerberus could see that I had not lied, and I came in, without having to drink from the Lethe. And so, due to such strange circumstances, I found myself there, and was going to speak about a world that, regretfully, I had not forgotten. It was quite a gathering. Only philosophers were present, all of them pre-Socratic, with an eagerness to know never seen before . . .

"What is the shape of the earth?" asked one of them. I replied that it is a sphere, and Parmenides rejoiced, while Heraclitus scowled. So many questions followed, in a pressing, quick succession, that I cannot recall them all. To Heraclitus, I replied that the universe is in perpetual change but that it had a beginning; to Anaximander, that our world is infinite, that humans were indeed born from other creatures, and that there is only one life in perpetual evolution; I described to Leucippus the atoms and their nesting of particles; Pythagoras was pleased to hear that numbers rule the world and that the laws of the physis are mathematical.

"Mathematician," he asked, "are you an initiate?" "Many of us are, presently," I heard myself shamelessly replying.

I don't know how long it lasted, and I felt weary. A long silence followed, full of meditations. Democritus was the first to speak. "Thus, with such a vast knowledge, humanity now possesses philosophy. Or am I mistaken?" he quickly added, catching a glimpse of my embarrassed look. I tried my best to appear confident and told them about the planet being invaded by technology, the population explosion, and the quest for values that would permit us to cope with an unprecedented situation. I saw some smile and many others frown. "How about the gods?" one of them asked. I did not reply.

At that point the interrogation began, and I found the place less welcoming. My judges made frequent pauses to consult each other heatedly, but kept relentlessly pressing me for answers. Each time they mentioned ethics I was at a loss, and they soon dropped the subject. "A barbarian," I heard, "perhaps even a slave or a craftsman."

I reacted. "Yes, a craftsman, as we all are now. We have been probing nature by means of experiments for four centuries, using the instruments of our craft as much if not more than our minds, and it is for this reason that we know so many things. If I tell you why we have difficulty understanding our own knowledge, perhaps you would be willing to help us.

"The sciences began among us as they did in Metapontum, when Pythagoras' disciples divided the immense curiosity of the master into so many branches of study. Some devoted themselves to music, others to mathematics, medicine, or plants; yet others took up the study of meteors or the substance of the world, and so forth. We are specialists, and that is our strength as well as our curse; even our philosophers are specialists." The eyes staring at me had no pity. I felt compelled to add, "But we are doing something about it."

"Thus," I went on, "at this very moment everything might be changing. Our experts communicate with each other, listen to each other; each of them is in turn master and pupil. It's as if the mind sought its unity. Our knowledge is so vast, and so many people are searching, that individual sciences merge. There is a quest for unexplored borders and so specialists come together, surprised to be in each other's company. Groups coming from different families are formed and, since necessity obliges, Agamemnon works side by side with Priam; such is the abundance of provender that they can feast together."

"Excellent!" someone exclaimed, and then added, "But why only now?"

"As you know well, you wise men, humans do not control their destiny, and if things happen, such as this coming together, it is due to the force of circumstances. It is taking place only now because an extraordinary event has occurred: we have just discovered that science is a whole. Don't laugh, perhaps you already knew it, but only instinctively, as a wish, while we are just leaving our divisions behind.

"Allow me to use an image to illustrate what has happened. Imagine reality (the universe, physis, whatever you may call it) divided up into plots, each of them the property of a particular science. Each individual science was busy digging, unearthing roots that it called laws, its own particular laws. At first, there was only a tangle of thin filaments. As the digging went on, the tiny roots began to join together to form thicker roots, reaching from one end of the field to the other. Soon they ignore cadastral lines and extend into neighboring plots. At present they form a harmonious lattice, no doubt incomplete, but without any gaps. No, Protagoras, this is not the consequence of the sole human will but something else: it is Reality, the Being, perhaps, structuring itself right in front of our eyes.

"Since when, you ask? It has been in the making for a long time. For more than sixty years now physics and chemistry have known that their foundations share the same laws; and it's not only yesterday that biology was invaded by chemistry and even by physics. But as for unity, which is harmony and cannot be reduced to a juxtaposition of components, it is only a generation ago that we have seen it dawn; on the mind's clock, this is yesterday, barely enough time for mentalities to change and take stock, for realizing where we stand."

"The One . . ." says Parmenides dreamily. "So it comes to you without being invoked or uttered. Lucky you mortals, for sharing this oneness and being able to seize it through the mind."

"Lucky the cities that possess philosophers," says someone who had probably just awakened, "for they have good laws."

He was made to shut up, but his remark only increased my uneasiness.

"Actually," I said, "that is precisely what we don't understand."

There was an outcry, interrupted by Democritus, who said, "How is that possible? At present you know, and therefore you have in your mind the clear idea, the exact image of things, as I did of the atoms. Hence, nothing would be easier than to communicate it with words. Isn't that what understanding, explaining the cosmos, is all about? What is holding you back?"

"Let me try to explain it using a remark of a friend of mine. 'Nothing,' he said, 'except laziness, prevents a physicist from understanding the leading ideas in biology; but for me, a biologist, nothing is more abstruse and obscure than the main ideas of contemporary physics or mathematics.' He was expressing what many others feel, and foremost among these, perhaps, the philosophers. If my friend is so interested in physics, it is because he believes that its laws are, in some sense, the closest to the essence of things (and no one mentions any possible reduction to another science). What is then his problem? It cannot be due to a different way of thinking, even though we sometimes hear references to the 'literary' and the 'scientific' types. Could it be that certain sciences, similar to music in that respect, can only be mastered at an early age, or that their study requires too much time? No, it is something else; and from listening to Democritus one may wonder whether physicists really understand their own science, or if they only have a long but superficial familiarity with it. They never have in their minds that absolutely clear image Democritus described; they may well have a partial one, a perception of overlapping fragments, of intuitive connections, but not a complete view."

I appeared to have offended Democritus. "Why, do you not see the atoms in your mind?"

"No. I try unsuccessfully to imagine them, but only mathematics can truly express the concepts and laws of the physis. Pythagoras will not find that idea surprising, but I'm afraid he would have said the same thing as you regarding mathematics and numbers: that understanding consists in having a clear idea in the mind, a sharp representation, and that the demonstration only serves to confirm its exactness. I also disagree with him on this point.

"I already mentioned a significant event that occurred only recently (the discovery of the unity of science) but another, less promising one, had preceded it—and almost prevented it, we might say. It was quite sudden, despite some foretelling signs, because it only took two quarter-centuries, one in the nineteenth and the other in the twentieth century. Three closely related sciences, logic, mathematics, and physics underwent a transformation almost at the same time. Without a common cause, all three moved from the visual, representable approach to the imageless, abstract, formal one. One can understand the case of logic, because it had always been formal without admitting it. Mathematics discovered that it does not deal with any particular objects but only with pure relationships, independent of any specific content, and so this science no longer has any maternal contact with reality. As for physics, it had once again to yield to the prevailing circumstances, if not to destiny. The more we penetrated the nature of space and time, and that of the atom, the more we realized that the only solid concepts we could use as a foundation were no longer 'visible,' 'expressible in words,' but of a purely mathematical nature."

"Are you saying," someone interrupted, "that for physics to move forward, to better reach the cosmos, it had to rely primarily on mathematics at the very moment the latter was breaking away from reality?"

"Yes, precisely. Although to be exact we should also mention experiments and intuition. Thus, we could say a great deal concerning the method of science, but some of us are so bewildered, or perhaps so perverse, that they do not believe there is a method any longer. Others maintain that science is a mere reflection of some spirit of the times, that it is completely transformed by revolutions, or that it is no more than a consensus among experts. How could a philosopher find his way through this conspiracy of abdications and ineptitudes?"

"Calm down," Parmenides the kind says to me, "these people are only too impatient to wait for the enigmas to be resolved at the appropriate time. Consider how long we had to wait ourselves. Tell us rather what your mathematics and physics are becoming, since they seem to preoccupy you particularly."

"Very well," I said, "it goes more or less like this. Our mathematics is presently entirely devoted to formalism, to symbol manipulation, to concepts that are constructed using axioms that defy any representation. Its structures are supported by a logic that is itself every bit as formal. Physics has found its primitive objects: space-time and the elementary particles of matter, but at the price of accepting that its principles and its foundations should be irrepresentable to the eyes of the mind."

I got carried away again and shouted, "Our sciences are blind like Homer, and like him, because blind, open to the entire cosmos."

"Then they are also mad like Homer, and like you, with your three-drachma lyricism," interrupted Heraclitus. "After all, they call me the Dark Philosopher, don't they? Why should there be only one way to understand? Have you considered that?"

"Yes, some have begun to consider it," I replied, hesitantly. "They wonder whether the obscurity of the founding sciences is inevitable, and if it is possible to 'understand' in ways other than the traditional ones. Certain recent works point in that direction. They concern quantum mechanics, a sort of science of the primal laws of matter, and the most haughtily formal. Those works were undertaken by their authors without any philosophical pretensions, only to clarify certain aspects of the theory. But they might have produced, unexpectedly, so to speak, something no one had searched for and that might well be a new way to understand, as you suggested."

"A science normally produces knowledge, but how can it also affect the nature of that knowledge and change the way we understand?" asked an impatient voice.

"That is true," I said, "we must carefully consider what is at stake from the point of view of a philosophy of knowledge. It is important to know that quantum mechanics rests on certain well-defined principles. They were discovered on the faith of certain experimental data, but their consequences have since been immense. They prompted a reformulation of the foundations of all physical sciences, and were many times confirmed in entirely new circumstances. These principles exhibit such a harmony that with them we can reach some previously unconquerable pillars. They are formal, though, as I have already told you; that is, the essential concepts involved are closer to mathematics than to anything our eyes can see or our imagination represent, such as wave functions—and there is still worse. The laws of physics are of course based on those concepts, and the properties of matter they express take the form of mathematical rules. No science could be more formal."

"Granted, but what's all that to do with philosophy?"

"You see, these new and well-established laws totally refute other principles of a philosophical nature, principles that had always been considered to be universal: intelligibility (the possibility of seeing what exists in space and time), locality (each thing is in some place), causality (there is no effect without cause), and a few others."

As I said this, I could see signs of concern appear on every face. Only Heraclitus seemed to rejoice. Democritus was stunned, and it was in a voice charged with emotion that he asked, "And what have you done to avoid that?"

"A great scientist, Niels Bohr, saw to it that order was restored. But he had to pay the price, a steep price. He didn't bring back common sense, or even the principles you consider as natural, but erected safety barriers beyond which thinking was forbidden. As a result, the philosophical aberrations we discussed earlier were prevented. Thus, he said, when we talk about atoms, we must refrain from asserting anything in regard to their position or motion; we can speak about them only using certain prescribed mathematical terms. Philosophers, of course, wanted to ignore those barriers, as did physicists, but all their efforts to cross the forbidden threshold ended up in failure. Some spoke of a veiled reality, to express this retraction of things under the scrutiny of the mind.

"A way around Bohr's restrictions could be found only recently. But even their authors did not initially understand what they had accomplished. In hindsight, they realized that the formal tools they had employed had provoked the equivalent of an epistemological earthquake, a true reversal in the order of knowledge."

"And so?" said a voice filled with impatience.

"The first step consists in reconsidering the roles of the laws and our perception of facts. We have always assumed that science proceeds from experience, from pure and visible acts easily translated into words, and from them it reaches the principles, not always transparent but nevertheless a synthetic summary of the facts. The new starting point is no longer experience, reality, but those same principles, which are considered to be more solid and certain than anything our eyes can perceive or our words express.

"Is that logical?" someone asked.

"Precisely. The whole question hinges on logic and its relation with physical reality; the problem of reason, if you wish. Why was Bohr forced to forbid thinking about the atomic world if not because the natural logic of language no longer applied in that domain? Some even thought that only a logic deprived of its roots could describe the external world. But one of the consequences of the new results was to show the existence of a convenient construction (which exploits the principles of the theory), thanks to which we can talk about the quantum world with an impeccable logic, albeit a formal one.

"Many questions become simpler by postulating a new principle concerning the use of logic in physics. As a result, some of Bohr's restrictions turn out to be irrelevant: precisely those that forbade us to understand. Basically, the idea is to understand differently: if logic has its roots in reality rather than in our mind, it is then possible to explain why our mind thinks the way it does. Today it is generally assumed that the mind is conditioned by our sensory perceptions: the world, from which our images come, and the components of our language. But the world we perceive is not the atomic world; it is made up of incomparably larger objects whose appearance, even if it originates in atoms, has completely different characteristics. If we can recover all the features of this macroscopic world from the most general and abstract principles of physics, then the reversal will be completed. Our vision of the world and the common sense that goes with it will no longer appear as a universally reliable starting point, but rather as a by-product of the laws of nature. As for the principles that were traditionally assigned to philosophy, we can demonstrate (for there are demonstrations) in which domains they are still valid. We shall not miss them, because we have better ones."

"In short," says Democritus, "science has given itself a unity. It had become obscure with the rise of formalism, and it becomes clear again by reversing the path toward knowledge. I once said that intelligence should come before knowledge, so the way things have turned out can only please me. Everything seems clear. Or am I wrong again?" he added, seeing that I refrained from approving.

"It is a peculiar situation. Nothing is yet certain, and we must proceed with caution because the possible consequences are too great."

"We have had enough of your convolutions! What is the problem?"

"There appears to be a gap, a chasm, between the world of thought, the theoretical world, and physical reality. It is as though the power of logic and mathematics, after accounting for the minutest details of this reality, were unable to penetrate its essence."

"So?"

"The theory is based more than ever on probabilities, on chance, because the possibility of a logical description of the world presently rests on this concept of probability. Thus, the essence of the theory is a description of what is possible, but the essence of reality is to be unique, so there is a gap between the two. We have perhaps reached the limits of what Husserl and Heidegger, an admirer of yours, called the Cartesian project: the theoretical explanation of the world using logic and mathematics."

I was interrupted by laughter of olympic proportions—it was Heraclitus'.

"Naive you have been," he said, "in believing that the irrepressible change could be confined within the perpetual immobility, the cosmos in the logos. Yet we have long been opposed to each other, he and I (pointing to Parmenides), and you thought it was purely by chance. But perhaps you are not as foolish as the Ephesians and, if donkeys prefer oats to gold, you have at least seen the bottom of the manger. The god whose oracle is at Delphi does not speak or conceal but he can make a sign and this is certainly one: to understand everything and finally come up against the ultimate limits of thought. You got there, but you complain, you lucky mortals; you don't realize how rich you are. Wipe your souls and give up your vulgar habits. The goal is there, within reach."

I heard Parmenides whisper to Zeno, "Do you think they are going to start philosophy all over again?" and the latter reply, "It would be a nice paradox."

PART ONE

THE LEGACY

✣

I F WE MUST RETHINK today the links between philosophy and science, it is because we are in the aftermath of a fracture. The most fundamental sciences, those dealing with space, time, and matter, those the Greeks would have called a science of Being, have broken out of the limits of common sense and traditional philosophy. Our objective is to identify, and in a certain sense repair, that void, that breach in the continuity of thought that prevents us from being fully aware of the state, the meaning, and the implications of science. The best way to begin is certainly by examining how such a situation was created.

It is convenient first to go back to the legacy, that is, to science when it was all clarity. Only then, against this background, may we be able to appreciate the evolution of knowledge in the course of time and see the turning of the tide. We shall then find the origins of what we might call the spontaneous epistemology of our time, a widespread conception of science that is persistent, short-sighted as well, at times fostered among philosophers by the writings of outdated authors—a conception stemming from yesterday's, not from today's, science. To rid ourselves of this conception we must recognize it for what it is, which means retracing the path that led to it. This is precisely what we are going to do.

The science we are going to talk about is not twentieth-century science, but that of the reassuring books that made everything look clear. It began with Bacon as a dream of philosophy, and it is not surprising if so many philosophers were inspired by it: Descartes, Malebranche, Spinoza, Leibniz, Hume, Kant, and many others. They fed it back to us in an elaborate form, making even more difficult the task of freeing ourselves from it.

To appreciate this dream, it suffices to remember the pre-Socratics, whose works, for the most part lost, all carried the same title, *On Nature*. We find precisely this *physis* in the writings of the Milesians, the Pythagoreans, the Eleatics, and those from Abdera, with their profusion of questions, naive and profound at the same time, like those of a child. How eager they were to learn why and how the sun shines, the sky is blue, the planets move, what are the elements, and how to pierce all the mysteries of life which they, our ancestors, could only marvel at! We have the answers, and we also know that true understanding requires more than that.

We shall talk of the time when science was young and still intuitive, in natural agreement with our perception of the world; classical, in short, as classical as Praxiteles' statues or Mozart's symphonies: a plain limpidity. We shall go right to the point with the help of a few significant examples, because completeness is not necessary. There are encyclopedias for that, and detailed knowledge may well hamper understanding.

We shall first talk about logic, the misunderstood. Like diamond, logic is pure, transparent, and also most impenetrable, capable of leaving its mark on everything. But we shall not discuss logic the way too many philosophy books do today, merely as a repetitious technique; rather, we shall explore parts of its history. Naturally, there is a purpose to all this, because the nature of logic ultimately raises the deepest question in science and philosophy, and we shall later try to unveil some of its mystery.

We shall also say something of mathematics. Why? Because of its central role in the structuring of the physical sciences. But mathematics will also have much to teach us by itself when, like a modern Logos, we shall see it slowly grow from a servant of science and philosophy to a would-be queen.

As for the sciences of the *physis* (nature), we shall restrict ourselves to physics, not from a mere predilection of the author but because it is the discipline that will reveal to us, later in the book, the major characteristics of contemporary science. Perhaps some readers will find our emphasis on the most basic parts of physics excessive. They might say, "No wonder one can get tangled with enigmas by following this kind of track. However, the truth remains that most of science is still clear, and becoming clearer every day, accessible to most people; we can see its mysteries being solved one by one." On this transparence of the flesh of science I agree, dear reader. Like you, I enjoy the new pictures of planets, the motion of continental plates, DNA molecules looking like an assemblage of balls, and all the rest. But then, behind the flesh there is the marrow: the laws, their fleeting significance, and their tantalizing unity. We are really only after that.

The first part of the book might well appear, however awkwardly, as a love song where the beloved is clarity. It must end in Schubertian melancholy, however, because the bride is taken away. Nothing shows it better than a glance at the history of philosophy, in the last chapter. At the time, philosophy too was living a period of enlightenment. The Greek inquisitive and deep questioning on Being had apparently been forgotten and everything was daylight and simplicity. For so short a time.

As we go along, there will be no conspicuous signs of a fracture, only slowly developing cracks. We must recognize them when they first appear, before they become chasms, and this is why we are going to revisit history in broad outlines. We have no other purpose in mind.

✣ CHAPTER I ✣

Classical Logic

LOGIC IS the daughter of Greece, as are democracy, tragedy, rhetoric, history, philosophy, and mathematics. It appears that in most earlier civilizations thoughts were uttered rather than constructed; truth was immediately recognized, not requiring any analysis to impose itself or be accepted. If humans have been thinking for a long time, it was only in a definite place and at a definite time that they began to dissect their own thought mechanisms in order to be able to reason. They were forced to admit that reasoning obeys its own laws, and that it does not give in to the will of the reasoner or the commands of the gods.

Logic has become for us the backbone of reason, even if we ignore what it is, as we ignore the justification for our almost blind trust in its power. When experts define logic as consisting of "principles for the validity of deductions," it is clear that they are attempting the impossible by using words without substance.[1] These basic questions about the nature of logic are nevertheless essential, for they will continue to bear upon everything we shall see afterward. The philosopher knows it, the scientist simply ignores it and carries on; it is the poet who says it best: "I'm but a maker of words. The words, who cares, and myself, who cares?" It is in a poem, rather than in one of his philosophical writings, that Nietzsche delivers this tragic confession.[2] Thus, every learned book is founded on ultimate ignorance. At the opening of this one, I would like to exorcise this curse. This is not an innocuous remark, for it

---

[1] See William and Martha Kneale, *The Development of Logic* (Oxford: Clarendon, 1978; 1st ed., 1962). This treatise has been our primary source.

[2] Saint John Perse, a French poet, said it magnificently: *O très grand arbre du langage et murmurant murmure d'aveugle-né dans les quinconces du savoir* (Oh language, standing like a high tree, you are also the mumbling whisper of one, blind from birth, wandering through the labyrinth of knowledge). It is probably impossible to translate. Poetry can convey the anguish we may sometimes experience regarding language and sense, because a poetic sentence is the exact opposite of a logical proposition. No word can be changed and many harmonious meanings sing together.

6

appears to imply that we cannot talk logically about logic in the absence of an already established foundation for language. There is no starting point for thought; it must begin with the imprecise, the conventional, the value of which will depend only on the extent of its fertility. Later on, this dark starting point will perhaps be enlightened by the knowledge that followed from it, and become part of a coherent circle. Such is our ultimate goal: to see the fruit of knowledge, born of an obscure seed, bear the same seed again with a meaning. Having said that, let us proceed without further questioning, since there is nothing else we can say on this matter for now.

To proceed means to accompany logic, in the present chapter, through its classical period up to the dawn of what was later to become formal logic.

Pythagoras and the Pariah

If I had to name the greatest thinker of all times, I would say without hesitation the unknown Pythagorean. After all, Pythagoras himself was perhaps only the one who came to announce the kingdom. We know that he was born in the Island of Samos, early in the sixth century B.C., and that he traveled to Egypt where he was taught by the priests of Amon, the human-headed god of Thebes. It is also said that he met the "naked philosophers" of India. He finally settled down in Croton, a Greek city in southern Italy, where he founded an ascetic and mystical sect.

He could have been just one of the countless gurus forgotten by history, and we are not interested in learning that he taught the transmigration of the soul, or that he was said to have a thigh made of gold. If we are interested in him, it is because of his presence, abundantly documented, at the origins of the intellectualism that was to impregnate Greek thinking. For Pythagoras, the intellect was the most important human faculty, one whose sole power can lead to a form of truth stronger and deeper than any other.

His vision of nature seems to us bold in the extreme. He said that numbers rule the world. This conviction appears to have been based on a simple fact: he had observed (or learned) that the lyre's harmonies depend on the exact place where the string is plucked, and that the musical intervals pleasant to the ear—octaves, thirds,

7

or fifths, as they are now called—come from strings whose lengths are in the same ratio as two whole numbers. However, to assert from this fact that "everything is number"—considered by some as the program of mathematical physics, even if it was stated well before the birth of either physics or mathematics—is an enormous extrapolation, almost utterly absurd, which leaves us stunned, in admiration, but also, we must confess, doubtful.

There exist many other examples of such astonishing illuminations among the pre-Socratic thinkers, often combined with ideas that are plainly fallacious. In fact, the genius of Pythagoras, and also of some of his disciples, was to have taken the first steps toward the *demonstration* of their ideas, that is to say, they knew how to *show* their ideas true in particular cases. To be sure, they did not entirely succeed but, as is often the case in the history of ideas, what they found turned out to be more important than what they were looking for.

Their first victory was the discovery of the famous Pythagoras' theorem for right triangles. Nobody knows how they did it, but most historians agree that they must have based their conclusion on some figure where the result may be immediately perceived by an attentive eye, and which does not necessitate any elaborate arguments. In other words, Pythagoras' theorem, just like that of Thales on parallel lines, is not enough evidence of a decisive progress in reasoning, and they only testify with certainty to a well-developed sense of observation. That theorem was most probably an observed truth, and not the result of unrelenting reasoning, but it was also an invitation to ponder over the mysterious number measuring the diagonal of a square, what we call the square root of 2. Which fraction was it?—for it could not be anything but a fraction made up of the only numbers worthy of ruling the world: the integers.

It is at this point that there enters the picture a man deserving the highest admiration and of whom we ignore almost everything, even his name. He was going to devote himself to the problem, no doubt after many others had done so. We may imagine him young, chosen by the Ancients for his brilliant intelligence when he was still an infant, a Greek child from southern Italy. I often dream of the unknown face of this hero of the mind. What bold impulse, caused perhaps by the failure of fruitless searches, or what compelling dream drove him to dare think the unthinkable: could it be that the elusive number had no name, that it was irreducible to the integers, those guardians of harmony? How to exorcise such a doubt?

We may presume that he had to meditate for a long time, treading as he was on uncharted territory. For the first time in human history, a man was going to establish an irrefutable truth by the force of reason alone. We ignore the particulars of his argument, but there are not many possibilities, and the records left by the mathematicians who were soon to follow him are conclusive. Proving that there is no quotient p/q of integers whose square equals 2 requires all the power of a logical argument. One must show that every even square is the square of an even number, and every odd square the square of an odd number; that one can always divide p and q repeatedly (both by the same number) until at least one of them becomes odd. One must especially be able to carry the argument to its successful conclusion, without leaving any loopholes, and to demonstrate that assuming the square root of 2 to be the quotient of two integers leads necessarily to a contradiction.

We can imagine the Ancients, unable to demolish his flawless argument, covering their faces with dust. He was cursed and declared blasphemous. According to one legend, the gods saw to it that justice was done in the form of a shipwreck. But it could have been the Ancients themselves who threw him into the sea in a broken-down ship, near the sharp reefs of the Calabrian coast.[3] Thus perished, perhaps, so that he would forever remain unknown, the one who brought us the light of reason, by Apollo anointed; Pythagoras, his forerunner, had been merely an omen.

He had opened a way, a boundless path, and it was now known that the mind, tightened by will and restrained by rigor, may have access to truth by the sole use of skillfully controlled speech. The mind had discovered its own strength, surprising itself. Logic was definitely born, with its inferences, its "hence" that stands no challenge, lest the challenger be swallowed by contradictions. At the same moment, mathematics too was born, because it was no longer limited to showing a property true by an example or a

[3] We only know with some certainty that a tomb for Hippasos of Metapontum was built while he was still alive ("Let him be declared dead!", meaning "We consider him as being already dead," and not "We want him to die"). All he had done was reveal the secret of the uncommensurables to the noninitiated.

figure, being now able to prove it conclusively by reasoning. Geometry was going to seize this brand-new tool right away and use it to create other wonders.

PLATO AND THE LOGOS

It is impossible to touch upon the theory of knowledge without first referring to Plato. He is not usually considered a logician, even if some of his dialogues contain several principles of logic. But his logical expertise is not systematic, and some of the rules he proposes are plainly wrong. He shows his talent elsewhere, in the *Theaetetus* and *The Sophist*, where he establishes himself as the first philosopher of logic by asking some fundamental questions that still mark out certain parts of today's science: What is truth and how do we recognize it? What is the nature of reason, and where does this faculty of deducing one truth from another come from? What is the nature of a definition, and what is the thing defined by the words? He attempts to provide answers to these questions but, despite their significance, we shall not discuss them, since their value is mostly historical. The context in which he places these questions is, on the other hand, much more interesting and deserves to be recalled.

Plato assumes the existence of "Forms" (sometimes translated as "Ideas," with a capital "I"), whose theory he develops in one of his latest dialogues, the *Republic*, with a strong Pythagorean flavor. It is easier to grasp the notion of form by resorting to examples and, rather than borrow Plato's—too dependent on their time—we shall use one from Descartes, which has the advantage of being very clear: "When I imagine a triangle, even if perhaps such a figure is nowhere in the world to be found except in my own mind, and it has never been, it does nevertheless exhibit a certain nature, or form, or definite essence of this figure, which is immutable and eternal, and which I have not created, and which does not depend upon my mind in any way whatsoever; as appears to be the case since one can demonstrate certain properties of this triangle."[4]

An Idea, in Plato's sense and such as described by Descartes, is not something concrete, something we could point at. A figure

---

[4] Descartes, *Méditations métaphysiques*, fifth Meditation.

drawn on a piece of paper is merely the image of *one* triangle, and not *the* triangle—essence of all possible figures of the same sort. Now, Plato never doubts the existence of an Idea of *the* triangle, something perfect, not of this world, and which is not simply the mental representation of a collection of figures, each one being nothing else than a particular idea deserving, at best, a lower case "i." The Idea is a "Form," that is, a perfect mold where the lowly ideas may dwell as so many interchangeable samples of their divine model. We shall quote two passages from the *Republic*, the first one stressing the uniqueness of the model to which its multiple manifestations conform: "Since the beautiful is the opposite of the ugly, they are two. And since they are two, each is one. And the same account is true of the just and the unjust, the good and the bad, and all the Forms. Each of them is itself one, but because they manifest themselves everywhere in association with actions, bodies, and one another, each of them appears to be many."

The second quotation illustrates well the nature of the problem that the theory of Forms intends to solve, which is to account for both the descriptive and the demonstrative power of language: "We customarily hypothesize a single Form in connection with each of the many things to which we apply the same name." We have therefore access to truth through reason *because* language refers directly to Forms, which have an independent existence and constitute the mold of all earthly things.

Forms do not belong to this world. They reside in a world of their own, an empyrean that Plato calls the Logos. To illustrate it, Plato turns to the famous myth of the cave: Humans are like prisoners chained from birth to the walls of a cave which represents the world down here. The real world, the true one, that of the Logos, is the external world full of light in front of the cave's entrance, where human beings move freely, there are trees, and animals passing by. The sun projects their shadows on the cave's wall, and the prisoners, seeing only these shadows, take them for the only reality.

It is therefore in the existence of the Idea that we must seek the power and the principle of the definition, which serves to liberate the unique Form from the variety of appearances and the multiplicity of manifestations. The faculty of reasoning, this possibility to demonstrate referred to by Descartes in the above quotation, results from the existence of certain particular Forms that are in communication with all the others, those expressed by words such as "being," "same," "other."

The theory of Ideas will be attacked by Aristotle, but it will reappear many times under different disguises. We know only too well the importance in theology of the idea of a divine kingdom, truer than the world of creation. Plato's conceptions will remain almost intact in the philosophical doctrine of realism, so popular during the Middle Ages, according to which words and ideas refer to Forms having their own reality, of a higher order than the reality perceived by our senses. The same Ideas can still be partially found today in what is called "mathematical realism," shared by the numerous mathematicians who, like Descartes, believe that mathematical concepts have an independent existence, of a different kind from that of the material world.

THE LOGIC OF ARISTOTLE AND OF CHRYSIPPUS

It is preferable to set aside for the time being the difficult questions raised by Plato, and go back to logic as a science and a method, in those days still looking for its own rules. The goal was not to determine the source of its power of persuasion, but a somewhat more modest and practical one: learning how to reason correctly, with enough caution to be protected against error.

From the outset, we can see opening up two different domains of application. One of these is mathematics, while the other, often tinted with rhetoric, aims at the correct use of the words and concepts of everyday language. Logic has always been torn between those two poles. The first domain, by its very nature and by its fertility, provides enough evidence of the power of logic, and it is in its deep relationship with mathematics that logic will finally find its purest form, albeit more than two thousand years later. On the other hand, the second domain—that of ordinary words and things—will not cease to remind it of the legion of traps into which the ambiguity of words or an incomplete knowledge of things may carry it, and it is in this realm that logic will first begin to purify itself.

Greek civilization has bequeathed us a sound logic, built over many centuries. Two different and often opposing schools contributed to its construction. The first in chronological order was that of Megaris, a city of Attica, on the Isthmus of Corinth. Its founder was Euclid of Megaris, not to be confused with the celebrated mathematician Euclid of Alexandria. Our Euclid was a contemporary of Plato and an heir to the Eleatic tradition emanating from Parmenides. The Megaris school would in turn give rise to the Portico (*stoas*) school or Stoicism, remarkable for the vigor of its research in logic, thanks in particular to the works of Chrysippus (281–205 B.C.). The other major school was that of the peripatetics, founded by Aristotle (384–322 B.C.).

We shall leave to the specialists the analysis of the differences and similarities between the two schools—which were eventually to converge, to a large extent. More important for us is to establish their joint contribution. This we shall do by staying as close as possible to the modern ideas whose origins we seek to determine— an approach certainly open to criticism.

It is well known that Aristotle considered reasoning by syllogism as the perfect archetype of logic. The example he used, transmitted to us through the centuries, is also familiar: "All men are mortal; Socrates is a man, hence Socrates is mortal." As a matter of fact, the syllogism does not really deserve all the attention it has attracted, for it leads to an unwieldy system of logic, long ago abandoned. A convincing example of a syllogism would be hard to find in any good mathematics textbook, ancient or recent.

The significance of Aristotle's analysis lies elsewhere, and foremost in the study of premises such as "Socrates is mortal," "A triangle has three sides," and so forth. He observes that these are not simple sentences but *propositions* that retain the same meaning regardless of their particular formulation. For instance, the sentence "Socrates is mortal" means the same thing as "Xanthippe's husband will one day cease to exist," not one word of which appears in the first version. Aristotle concludes that, if logic seems inseparable from language, it lies at a higher (or at least different) structural level, that is, in the domain of meaning we call semantics.

It is not always easy to tell language apart from semantics, or a sentence from a proposition, and logic will often get entangled in such obstacles. Indeed, words can have a thousand meanings, a thousand connotations, and when we say, for instance, that "Socrates is a rose," it is not at all obvious that we have not uttered a proposition, for comparing someone to a rose admits many symbolic interpretations. This initial difficulty was only going to be resolved by the formal logic of our time and its notion of "universe of discourse," which amounts to carefully restricting a priori the propositions we are allowed to consider.

Propositions are the pawns logic moves forward, those that it conjoins, compares, opposes, and combines to create new ones. How does it do it? Aristotle and Euclid of Megaris note that propositions may take two forms, at the same time different and inseparable, one of which is positive and the other—its contrary—negative; for example, "Socrates is mortal" and "Socrates is not mortal." Logic does not restrict itself to finding and telling the truth, as an oracle would, but it initially places on the same level the eventually true and the eventually false before reaching a decision. This is based on a fundamental rule that we owe to Aristotle, the *principle* (or *law*) of the excluded middle: a proposition must be either true or false. Even today, this principle is the cornerstone of logic, and anything having the appearance of a proposition but not obeying the principle must be banned from the garden of logic.

Aristotle is also breaking new ground when he distinguishes between universal propositions ("Every living man has a head") and particular propositions ("Some men are red-haired"), and he clearly indicates the difference. Modern mathematical logic has even introduced specific symbols for each of these forms, which are stated beginning with the standard "for all" (or "all") in a universal proposition, and "there exist" in the particular ones. Thus, the above examples would become "All living men have heads" and "There exist red-haired men."

We shall not accompany Aristotle any further, and rather follow the works of the Stoics, in particular those of Chrysippus. It is worth noting that it was Chrysippus who Clement of Alexandria used to mention as the master of logic, together with others, such as Homer in poetry, Aristotle in science, and Plato in philosophy.

Rather than using syllogisms, which soon become cumbersome as the number of premises increases, Chrysippus calls attention to some simple and better ways of combining propositions. It suffices to wisely employ the short words "or," "and." He specifically distinguishes the exclusive "or" from the nonexclusive "or," the first one corresponding more closely to "either, or" ("Either you buy the newspaper or you put it back on the shelf"), while the second allows for several possibilities not necessarily incompatible ("I enjoy reading novels or funny books," which is nonexclusive, for some novels may be funny).

Chrysippus managed to find the proper rules to manipulate what are now called the logical functions "and," "or," "not." They are so named because, just as mathematical functions do, they associate a well-defined object to one or several given objects—propositions, in the present case. Given a proposition a, the function "not" defines another proposition "not-a"; in the same way, given two propositions (a, b), one can form a new proposition, "a and b"; and similarly for "or." Chrysippus not only identified the connectives but gave precise rules concerning the composite propositions, such as "a and a" = a; "a and not-a" is impossible (this is the law of the excluded middle). There are more than a dozen rules we probably owe to Chrysippus, although it is hard to differentiate his contribution from that of his successors. Let us notice, in passing, that the use of letters to represent propositions as we have just done, and Aristotle and Chrysippus also did, was common practice among the Greek.

The important notion of deduction, also called logical inference or implication, was recognized and clarified as well. It comes up in sentences such as "If a, then b," usually denoted by $a \Rightarrow b$. Deduction is without question of primary importance in logic, for it is thanks to it that we can build arguments leading from hypotheses to conclusions. Two rules of great significance also appear at this time: transitivity, according to which $a \Rightarrow b$ and $b \Rightarrow c$ entail $a \Rightarrow c$; and reciprocity, which decrees that the conditionals $a \Rightarrow b$ and not-$b \Rightarrow$ not-a are equivalent. Finally, the nature of the initial truths is elucidated. These are propositions whose truth is assumed from the beginning, either because it is self-evident (the axioms) or is accepted by convention (the postulates).

On the whole, the essentials of logic have already been conquered before the end of antiquity. If anything, logic contains too much, too many outgrowths which do not really belong to it but result from the fact that the development of the physical sciences is trailing that of the science of reasoning. Also, a considerable amount of the logical expertise of the Stoics will be ignored or misunderstood for a long time, because imperfectly transmitted during the Middle Ages, and systematically underestimated in favor of Aristotle's (and his commentators') works. That ancient knowledge was again overlooked in the modern age, our civilization having clearly fallen behind in the domain of logic until its revival during the nineteenth century.

Let us sum up the main traits of this logic, to be used occasionally later on: First, it is essential to delimit a *domain of propositions**⁵ or a domain of thought (*Denkbereich* in German). These propositions must clearly obey the law of the excluded middle. Then come the axioms. They can be self-evident truths, principles, or simply hypotheses. Propositions give rise to new ones through the use of the logical functions "and," "or," "not"; the truth (or falsity) of the latter depending on whether it can be established by deduction from the (assumed) truth of the axioms.

The undisputed masterpiece of ancient logic remains Euclid's *Elements*, manifestly written with little influence from Chrysippus, despite the fact that the mathematician and the logician were contemporaries (but the former lived in Alexandria and the latter in Athens). Logic proper appears less clear than mathematics for it repeatedly fails to deliver, employed as it was to treat foggy, inscrutable subjects: nature and the gods.

THE PARADOXES

For all practical purposes, the history of what we have called classical logic ends in the third century B.C. The sap has dried out. To be sure, logic would be revived in the Middle Ages by Scholastic philosophy, but without adding anything of substance to what was already known—on the contrary, as we have just seen, the meaning of the Stoics ideas would be partially lost. The Renaissance and the classical period marked, surprisingly, a regression. The famous *Logique* of Port-Royal, by Arnauld and Nicole, does not measure up to the medieval works of Albertus Magnus and William of Okham. This temporary regression may be explained by the development of science. Instead of proceeding by pure reasoning and from postulates that were often arbitrary, science found a new im-

⁵ We have tried to restrict the use of technical terms, but some of them are nonetheless convenient, even if not always familiar to the reader. A brief glossary appears at the end of the book. An asterisk indicates the first occurrence of a term on that list.

petus through observation and experimentation. Very few thinkers are then watching over logic's unsteady flame; only Leibniz, among the truly great. Perhaps that was precisely what enabled science to be born, by shedding the burden of intellectualism and its illusory dreams. In fact, logic would only reappear in the nineteenth century, under the pressure of new and difficult questions coming from mathematics.

We will steer clear of the main trends in logic, which remained dormant for a long time, and simply glean some grains of wisdom from antiquity. As we have already mentioned, it is important to consider only propositions that satisfy the law of the excluded middle. Violating this condition, which is not always easy to verify, may lead to paradoxes. Etymologically, a "paradox" is a proposition that seems to say something opposite to common sense, but the word is gradually replacing what used to be called an aporia (the precision so dear to logicians sometimes borders on pedantry), that is, an untenable proposition, often self-contradictory.

In the Megaris school, paradoxes were gladly exchanged, often in a playful mood, as in the following example involving "horned," a word with undertones of conjugal infidelity. It began with the premise "What you haven't lost, you still have." The naive conceded that much, only to be told, "You haven't lost your horns, hence you still have them." Joyous laughter followed on the squares of Megaris. Just a joke, you may be thinking, except that some of Plato's own arguments, supposedly serious, were not much better. It was the time when logic was trying to find its way, and paradoxes taught it how to protect itself against its own traps.

The ancestor of paradoxes dates from more ancient times. It is due to Zeno of Elea, a pupil of Parmenides and older than Euclid of Megaris. Zeno wanted to defend Parmenides' claim that "The Being is immobile" against some genuine objections borrowed from Heraclitus and other less serious criticism dictated by common sense. Indeed, they said, Parmenides' proposition is absurd, for everything is in motion, including the celestial spheres, and nowhere in this world is there room for the eternally immobile. That's a mistake and an illusion, replied Zeno; motion does not exist, because it contradicts itself. Here is my proof: Can Achilles with his winged feet reach the boundary stone of the stadium? He needs a lapse of time to travel half the distance, and still another lapse of time to travel half the remaining distance, and so on and so forth. Hence, he needs infinitely many time intervals to reach the stone; but that is an infinite amount of time, as you will all agree. And so Zeno stopped Achilles, "immobile at great strides," by means of words alone.

We are no longer troubled by this paradox, because we know that the sum of an infinite number of (unequal) time intervals may be finite. This example is nevertheless interesting, because it reminds us of the extent to which the logical treatment of infinity is subtle. Thomas Aquinas himself was tricked by it into error, and it is thanks to infinity that logic will have the opportunity to be reborn, late in the nineteenth century.

Let us mention a last paradox from Megaris, one that is still very popular: the paradox of the liar. We understand here by "liar" not just someone who habitually tells lies, but a person who never tells the truth. The most familiar version of the paradox goes like this: "Epimenides, the Cretan, says that all Cretans are liars." This is clearly a paradox: if Epimenides tells the truth, he is an example of a Cretan who has told a truth, hence, he has lied. If he lies, the contrary of what he says—Cretans never lie—must be true, so he must be telling the truth.

Rather than a paradox, this shows how one can play on the meaning of the words. Indeed, the negation of "All Cretans are liars" is "Some Cretans tell (sometimes) the truth"—and not "Cretans never lie." There is therefore a way out. But how about the man who declares, "I am lying"; either he is telling the truth, and then he must be lying, or else he is lying, in which case he tells the truth. This is already more difficult to explain away, and we can see that what is called into question here is the law of the excluded middle.

Modern logic would split the problem in two. Propositions of the type "X says that . . . " were studied by the Anglo-Saxon philosophers of language, logicians generally believing that they do not belong to the field of logic. But there is another angle to Epimenides' example: an element (Epimenides) of a set (all Cretans) occurs in a proposition that refers to the whole set. Logicians recognized the prime significance of this aspect, and that they ought to exercise the utmost care when employing the word "all."

The two lessons to be learned from all this are that we must be careful not to succumb to the absurd when dealing with infinity, and that the same applies to totality.

Two Useful Notions

In what follows, we will have the occasion to put to use two notions from the domain of logic. The first notion is a basic one, and it is usually referred to by its Latin name: *modus ponens*. The second notion, which belongs rather to the philosophy of logic, concerns "Okham's razor."

*Modus ponens** is a topic in pure logic. Although explicitly formulated by Abélard (1079–1142), it was already known to the ancients, since Euclid (the mathematician) systematically employed it to prove new theorems from old ones without having to go back each time to the initial axioms and postulates. In everyday life, all sorts of people—engineers, technicians, researchers, teachers, and students—use *modus ponens* every time they use a theorem or a formula whose proof they do not remember in detail. We have essentially the same thing in logic: the possibility of starting, in the middle of an argument, from a proposition previously established, without having to justify how it was proved. Modern logicians, who are careful not to sweep anything under the rug, have demonstrated the soundness of *modus ponens*. We will leave it at that for the moment.

"Okham's razor" is more like a guiding principle for thought, capable of shaving in many other domains besides philosophy or logic. I mention it now but will not use it until the very end of the book. William of Okham, of whom we know only the year of his death, 1349 or 1350, was a Franciscan. He could have been the model chosen by Umberto Eco for the hero of his novel *The Name of the Rose*: a sensitive man, a sharp intellect, and a prolific author. He is better known in the streets of Oxford for the following rule, Okham's razor: *Entia non sunt multiplicanda sine necessitate* ("Entities are not to be multiplied without need," or why use many if few will do? do not imagine multiple causes where only one is enough, always try to keep the number of your hypotheses to a minimum, define the domain of your discourse as precisely as possible). In logic, do not multiply the number of axioms, and eliminate redundancies, as Euclid did in his books. Do not hesitate to apply the same principle in metaphysics: when you refer to God as the Creator, it is pointless to assume other attributes of creation because they are already present in the nature of God.

Conversely, if grace fails you or intuition does not reveal to you what this divine nature is, do not encumber your reflections on the terrestrial nature with your ideas about God. Do likewise in philosophy and in science, by reducing the number of principles. Clarity will ensue.

THE UNIVERSALS

We shall close this chapter with an important page in the history of logic, written in the Middle Ages. It is no doubt the only historical example when a question concerning logic was passionately debated, provoking endless public controversies and the intervention of kings, popes, and saints. This is precisely what happened in the eleventh century, when the small community of students and clerics got all excited about a philosophical contest opposing the great intellectuals of the time. Some of them are known outside scholarly circles even today. Who has not heard of Abélard, great master of seduction, who knew how to stir the enthusiasm of boisterous students, eager for a resurging knowledge? Who does not know Saint Bernard, the preacher of the crusade and the rebuilder of monastic life, whose passionate and mystic personality placed him, with their concurrence, above popes and kings? The controversy opposed these two men, as well as countless others after them, for it would last almost two centuries.

This nominalism-realism controversy, as it is called, concerns a question at the center of the philosophy of logic, important enough to spill over the whole framework, and even the nature, of philosophy. It is the question of the value of language as a means for attaining truth, or, in other words, the foundations of the theory of knowledge. As Bertrand Russell rightly observed in his *History of Western Philosophy*, this dispute raised a question that remains as relevant as ever and at the heart of contemporary thought.

The question's original formulation was both more precise and narrower than Russell was later to make out. It involved philosophy as it was taught at the time, under the combined influence of Aristotle and Plato. What is the nature of the "universals"?—a term that has practically disappeared from our language and which used to designate the concepts associated with words. A universal is therefore a generic name such as "man," "kindness," "animal," "soul," "being"; all part of the vocabulary of philosophy in its quest for knowledge. Knowledge progresses by an analysis involving the judicious use of words; its conclusions are expressed in words, and, in the Middle Ages, its sole mode of inquiry was speech, the endless combination of words. It is therefore essential, prior to any development of philosophy, to agree on the meaning and the role of language, and in particular on the nature of the universals. We must not forget that philosophy's primary purpose was to serve as the basis for theology, the latter being but a scholarly commentary on a divine message, at the same time revealed and clouded by the words that carry it—but we shall not insist on this particular aspect.

Two main theses clash from the beginning. We shall attempt to summarize them without any claim to completeness, nor shall we try to follow their evolution over time. The first position is that of the supporters of realism. It is the great Platonic theory in which the Ideas (or universals) are real. Medieval realists, however, do not go as far as Plato in claiming that Ideas are more real than material reality; nevertheless, they believe that they are conceived by God for all eternity. The opposite point of view is nominalism. This second thesis would profit the most from the dispute, and become more elaborated as the latter develops. It originally appeared in a form so transparent that it almost mocked itself: general concepts are nothing more than the resonant utterings made by the mouth in pronouncing the words; mere sounds that we use to describe, in a more or less arbitrary manner, what we observe. Or, as pointed out by Roscelinus—one of the first to enter the dispute—they are only swellings of the voice.

There were no clear winners of the controversy, and interpretations vary according to the sources, the Dominicans (notably Albertus Magnus and Thomas Aquinas) or the Franciscans (with Duns Scotus and William of Okham). On the whole, a moderate form of nominalism carried the day. The universals are modeled on the reality accessible to humans (which includes part of the divine reality). This reality exhibits a certain order, which results in similarities within what we call, depending on its degree of generality, a genus (for instance, tree, stone, or human) or a species (oak, ruby or lustful). However, the human mind has, to a large extent, the privilege to choose as it pleases the criteria and the borders of the categories that it decides to name.

A short lapse of time separates the great masters of the end of scholasticism from the first shudderings of science. The focus will then be on the order present in nature, which is at the origin of the practical and the semantic applications of the universals. During the Renaissance, a research method was even developed consisting in the comparison of the words used to designate natural events. Whatever the case, it is only with Locke and Hume that this question in the semantics of logic will be rekindled. We shall meet it again then.

Classical Physics

THERE WAS A TIME when things seemed to be really as we perceive them, and physics was then "classical"—natural, simple, we may also say, or even naive, had it not soon become too rich to fit any of those descriptions. This youth of science will be our next topic, from its origins till near the end of the nineteenth century. Of course, we shall not retrace its entire history, only put out enough landmarks here and there to see how the rise toward formalism gradually imposed itself and, at the same time, consistency settled down. Maxwell's electrodynamics marked the end of this age of innocence. Afterward, nothing was ever the same.

ASTRONOMY, FROM HIPPARCHUS TO KEPLER

Would science have ever been born, wondered Henri Poincaré, if man had not been able to contemplate the stars' peaceful and ordered parade through the skies? Wouldn't eternal clouds, such as those covering the sky of Venus, have darkened the mind as well as the heart? As for sunshine, who knows what craving for purity and brightness it can inspire? Babylonians, Chinese, Indians, Egyptians, and Incas kept records of the heavens; and the northern peoples too, from Stonehenge to the Mongolians, those worshippers of the eternal blue, tracked the constellations and their swaying across the sky at the beat of the seasons.

Mathematics' first stammerings appear to have been related to the observation of the sky. The prevailing regularity of the celestial bodies was perhaps an invitation to confirm it, to make it explicit, and to predict it by means of numbers. Among the western civilizations, the Babylonians did it, and the Greek intellectuals will do it too, once equipped with real mathematics. It took a shrewd mind to discover, early on, that the earth is round, as its shadow on the moon indicates (Parmenides is credited with being the first), and later measure quite accurately its circumference (Eratosthenes,

284–192 B.C.). A few years earlier Aristarchus of Samos had already estimated the distance from the earth to the sun and the moon.

All these discoveries were not exclusively motivated by a yearning for knowledge and understanding. They were rooted in a preexisting representation of the world. The desire to predict the march of the planets was intimately connected to a very ancient belief in their influence on the life of men and empires. In Pythagorean intellectualism, among other schools, the celestial world was inexorably coupled with an idea of perfection. This association will push Aristotle to conceive principles that are purely mystic: the paths of the celestial bodies must be perfect, hence they can only follow the sole perfect curve, the circle (the circle's perfection was justified by its being the only curve equal to itself at every point). Aristarchus provides another example of this difficulty in abandoning the traditional representation of the world. Hadn't he proposed that the celestial phenomena could be more easily understood by assuming the earth to be simply a heavenly body moving around the sun? But then the earth would be carrying along with it the Olympus, the abode of the gods. What a sacrilege! This impiety would cost Aristarchus dearly. He was condemned and had to renounce his idea, or at least keep it to himself.

The paradigm of the Greek astronomer was Hipparchus. He lived in the second century B.C. but no one can tell exactly when, a cruel irony for a man of numbers and a master of time. Like all his predecessors, he believed that the stars are fixed to a celestial vault, a spherical canopy of heaven revolving around the earth in twenty-four-hour cycles. Each star thus describes a circle, the perfect curve. Hipparchus keeps a detailed record of his observations, tracking the precise location of heavenly bodies in the course of time. He also makes use of ancient data, and will eventually discover the precession of the equinoxes (the occurrence of the equinoxes earlier in each successive sidereal year), which he interprets as a slow swinging of the axis of the stellar sphere.

He also notes a discrepancy between the actual planetary motions—counting the sun and the moon as planets—and the circular paths that perfect bodies would be expected to follow. The planets are therefore only almost perfect, as their close proximity to the earth suggests. Hipparchus then wonders what kind of motion, less than perfect but still appropriate to celestial bodies, might animate them. To his astonishment, he finds two possible answers. The first one involves the combined motions of two circles, C_1 and C_2, say. The center of C_2 travels along C_1 with uniform circular motion, while at the same time the planet describes the circle C_2, also with uniform motion. The resulting planetary path is a rather complicated curve, the epicycloid (from *epi*, "on top," and *kuklos*, "circle"). The second solution is that of the "eccentrics." It is conceivable that the moon, for instance, should indeed describe a circle (with uniform motion), but one whose center is different from the center of the earth. The existence of these two solutions (neither of which, by the way, is entirely correct) will play a central role in the history of philosophy, and will provoke some of the oldest and most profound reflections on what it means to understand the world. We shall return to it.

After Hipparchus, as the observations became more accurate and took place over longer periods of time, sky watchers realized that neither epicycloids nor eccentrics could account for the motions of Mars and Jupiter. It then became necessary to resort to more complex constructions, involving three or more circles rolling over each other, and giving rise to even more intricate trajectories, the epicycles. The necessary calculations, extremely difficult considering the available means, were performed principally by Ptolemy of Alexandria (90–168); their accuracy in predicting eclipses, conjunctions, and oppositions was truly remarkable.

Our purpose not being to relate the history of astronomy, we shall skip the valuable Chinese observations as well as the medieval works of Arabic and Persian astronomers, and move on right away to Copernicus (1473–1543). Shortly before his death, he published a work summarizing his calculations of the celestial motions over many years. These calculations are based on Aristarchus' hypothesis, by then forgotten or simply ignored: the sun, not the earth, is the center of the world, and the latter rotates around the former. Planetary motions are still explained in terms of epicycles, but they become considerably simplified. For instance, in the new theory, the apparent motion of Jupiter as seen from the earth results from the combination of two motions: those of the earth and Jupiter each rotating around the sun. From the apparent motion of the sun he infers the motion of the earth, and then uses the latter to systematically correct the apparent motion of each planet. In so doing, Copernicus greatly simplifies the system of epicycles.

Much has been written about this "Copernican revolution," which offers two very different aspects. The first one, purely empirical, is a notable but rather technical progress that only the experts could appreciate: the number of epicycles was reduced, thus simplifying the calculations—which were in any case the business of a small number of people. The second aspect is an unprecedented event in history: humankind has changed its representation of the world in a space of one generation.

Rather than repeat things that have been said one hundred times before, we shall refer to Giordano Bruno (1548–1600) to demonstrate what was at stake. His case, by its extreme character, constitutes the best example. He was a well-educated man, being a Dominican—at least until his daring views got him expelled from the order—and driven by a strong desire to understand. Copernicus' theory was for him like a second Revelation. From it follows that the earth is merely a planet, and the sun just another bright celestial body, no different from any other star. Hence, there is no particular reason for this sun to be at the center of the world; this center is everywhere: the universe is infinite. There are also countless stars, separated by enormous distances, as proven by the feeble light that reaches us; and around each one of them there must be other planets, no doubt inhabited, just like ours. Bruno is not really a physicist. His modest contribution to this science was limited to some relevant remarks on the centrifugal force and on the earth dragging along the atmosphere, by which he explained our being unaware of the earth's rotation. If he seems great to us, it is as a theologian and philosopher: he dares turn the Thomist method against itself, drawing the boldest conclusions from the new ideas, and undermining the most sacred dogmas to reach a pantheistic vision of the world, where creation and Creator become one and the same entity. We know that his views cost him his life by fire, victim of a vision of the world that nobody wished to see changed.

Let us close this parenthesis and move forward to Tycho Brahe (1546–1601), the perfect model of a sky watcher. A Dane, member of the nobility, he had a number of measuring instruments built for his observatory in the island of Uraniborg, astrolabes and gnomons of the highest quality, although none of course equipped with optical devices. He observed the sky for more than twenty years, recording the position of heavenly bodies and events in his Rodolphine Tables, to be completed at Ratisbon. It was there that he hired as his assistant a young German with a gift for calculations, Johannes Kepler (1571–1630).

After the master observer, here comes the master theoretician. He deserves more than just a few lines. We have already exposed the strange detours that led to the theoretical basis of Greek astronomy: First, there is the visual reality, the heavenly bodies, and the dream, perfection. The ideal is to bring them together. This seems attainable, but there are discrepancies, forcing successive modifications of the original idea until there is practically nothing left but an empty tradition. In Kepler's time, the initial dream has run its course and can now be abandoned. It has nevertheless left a concrete trace: the long and tortuous calculations that had managed to chart the heavenly motions. The present situation is fluid—which perhaps partly explains Kepler's whimsical personality—somewhere between yesterday's failure and tomorrow's hesitant promise. Whatever the case, he will set out to impose a mathematical order on Brahe's mass of information, using as a guide the numbers themselves, rather than metaphysical preconceptions.

Kepler is one of those tormented men, in constant pursuit of harmony, who made the Renaissance. He ponders over Tycho Brahe's accumulated data as if trying to solve a puzzle, striving to detect a hidden order that he will slowly uncover. First comes the law of areas, in 1604: the line segment connecting the sun and a planet sweeps out equal areas in equal times. One year later he formulates a new hypothesis on planetary motions, not the first one to be put to the test: planetary orbits are ellipses, with the sun located at one of the foci. The testing of each hypothesis then necessitated a tremendous amount of difficult calculations. It does not take much, though—an accidental juxtaposition of numbers, perhaps—to suggest a new, hitherto unexplored possibility, as anyone who has had some experience with complex calculations knows well. Thus, we should not be overly surprised, or look for some profound reason (most likely suggested by hindsight) for the fact that a geometrical hypothesis so unexpectedly simple came up in Kepler's calculations, for other calculations had preceded it. What is truly new is his stubbornness in searching for some kind of order at all costs. This time, the data fit the hypotheses marvellously well, and the epicycles are dumped forever. Finally, in 1618, Kepler discovers a third pattern in the planetary motions of our solar system: the cubes of the major axes of the (elliptic) orbits are proportional to the squares of the planetary years (the time a planet takes to complete one revolution around the sun).

A novel idea will then slowly begin to ripen: would it be possible that lifeless nature should obey an order imposed by mathematics? In fact, the idea goes back to Pythagoras, but its present form is a sort of converse. For it is no longer a question of beginning with preconceived ideas of perfection, formulated in mathematical terms, and then forcing them on the facts. Quite the contrary. Now one starts with the bare facts and then tries to see whether they structure themselves according to some mathematical rules. Such rules are in a sense empirical, for we accept them without necessarily understanding their profound significance. But finding these rules might often require a fertile imagination and considerable effort, as Kepler's own case shows well. His three laws are the paradigm, cited over and over again, of this notion of empirical rule.

We are by now so used to seeing material reality accommodate itself to numerical rules that it is at times difficult to appreciate the astonishing fact that such rules should exist at all. More astonishing is the almost certain success we encounter whenever we set out to look for one and, an even greater wonder, how harmoniously all these rules fit together instead of contradicting one another. With Kepler, astronomy has fulfilled its role of midwife in the birth of science, by revealing the existence of empirical laws shaped in mathematical form.

THE DAWN OF MECHANICS

The origins of mechanics are fascinating in their simplicity, and they are proof that the concepts of a science can be derived from the most routine and everyday experiences. The resulting representation of the world not only fully agrees with our intuition but actually completes it. Just as Poincaré wondered whether humankind would have discovered science without the view of the nocturnal sky, we can ask ourselves whether that discovery would ever have been possible without this continuity between the ordinary and the scientific, a continuity that we have lost since. There is no better opening for this new chapter of history than Einstein's famous statement: "The Lord may be subtle, but He is not wicked."

This patent simplicity can be found with particular limpidity in the notion of force. The concept appears already in the most ancient times, for everyone knows what it takes to lift a weight, to haul a chariot, or to bend a bow. The weight provides a means to measure a force, thanks to the balance. Archimedes (278–212 B.C.) seems to have been the first to draw attention to the significance of the point where a force is exerted, an essential element of his theory of levers.

Many centuries later, Stevin, also known as Simon of Bruges (1548–1620), would state the laws of the balancing of forces acting on a body at rest—in other words, statics. With the aid of ropes, levers, and pulleys, he demonstrated beyond doubt that a force is completely characterized by its magnitude, its direction, and its point of application, for only these parameters enter into the equilibrium laws. He also showed how several combined forces have the effect of a single one, the resultant force. He computed it using the "method of the parallelogram," which gave rise to the modern notion of addition of vectors. So not only physical concepts, but also occasionally mathematical ones, have an empirical root. The above example shows that the new knowledge points toward a much more mysterious concurrence between physics and mathematics.

Once the principles of statics were understood, there remained the question of dynamics, that is, the relationship between force and motion. The ancients had of course already noticed that a force can generate movement: a horse pulling a cart sets it in motion. They also believed, as Aristotle did, that the inverse was also valid: motion would last only as long as there was a driving force. What can we say about this "evidence" but that logic may be deceptive? We know the sequel: there must be a continuous force keeping the arrow in its course. Modern authors remind us of the scholastic solution: this force is exerted by an angel. We may smile at this "solution," but it had better be a large smile, for the same kind of explanation will reappear later—remember the ether—and perhaps even today, with the somewhat mystical properties of vacuum in quantum field theory.

The question continues its slow progress through the Middle Ages, until it is settled by Galileo (1564–1642), for whom it is clear that motion may exist in the absence of force. Not an entirely new idea, though, since it can be found in the works of Oresme (1320–1382), but Galileo's crucial contribution will be his systematic application of the experimental method. He studies the motion of a ball on a horizontal plane. When the ball is at rest, statics tell us that no horizontal force acts on it. Galileo assumes that the same is true even when the ball is moving, and his observations confirm it: the ball travels in a straight line with constant velocity, provided friction does not slow it down. This is the origin of the *principle of inertia*,* which will play a central role in the history of physics: a body on which no forces act will travel in a straight line with constant velocity. Actually, the exact formulation of the principle took some time to crystallize, and the one we have just given is not due to Galileo but to Descartes. Galileo believed the motion to be circular, corresponding to the rotation of the earth, rather than linear—but never mind.

We know that Galileo also studied falling bodies, using once again the experimental method. He rolled balls on inclined troughs so as to reduce the effect of weight; the resulting slower motion was then easier to measure. His conclusions being well known, we shall only underline their simplicity, which confirmed his famous creed: "The Book of Nature is written in mathematical language."

To these initial laws of dynamics, Huygens and Wallis added (around 1670) those concerning collisions, where the concept of mass, by then clearly distinguished from that of weight, plays an essential role. Two new quantities enter the picture: momentum and the *vis viva*, what is now kinetic energy. All these "laws" are essentially only empirical rules, simpler than those formulated by Kepler.

We must still mention one more tool, essential to dynamics: analytic geometry, invented by Descartes in 1637. Basically, it reduced geometry to algebraic calculations on the *coordinates* of a point, that is, on the three numbers that locate the point's position with respect to three axes, the reference system. Euclidean geometry was perfectly adequate for the study of certain specific curves, conics and some others, such as the famous cycloid, which so delighted the mathematicians of the time. But this geometry was rather impractical and often useless for describing, or even imagining, more complex trajectories. By reducing such descriptions to computations, Descartes provided a precious new tool. Now each coordinate could be considered a function of time whose exact form was to be determined by the theory.

As a sidelight on history, let us observe that Newton hated everything coming from Descartes. For him it was therefore a question of honor never to use Descartes' method. He could actually manage without it, because the most important problems he had to face involved only trajectories that were conic sections. And so, in his great book, no mention of Descartes became necessary. But his successors soon ignored this interdiction, which Newton had been careful never to state explicitly. We may also see things from a different angle, more anticipatory of the future: space plays in dynamics the role of a physical container, and so the possibility of describing it abstractly using algebra is perhaps the first sign, still uncertain, that formal science had arrived. But how could anyone see it yet?

NEWTON'S DYNAMICS

The works of Newton (1642–1727) in dynamics remain unquestionably one of the pinnacles of science, never surpassed, even if other achievements may pretend to match it. He published them in 1687, in his *Philosophiae Naturalis Principia Mathematica* [*Mathematical Principles of Natural Philosophy*], the bulk of which he said he had worked out during his youth.

Among the multiple aspects of his genius we shall emphasize the totally new twist he gave to the "laws" of physics. Before him, these laws appeared merely as empirical rules, extracted after a careful analysis from the mass of facts. But Newton introduced "principles," universal laws that nature obeys, and from which the former empirical laws follow as logical, mathematical consequences. This supposes in particular that we must, at least in thought, rid ourselves of our terrestrial condition and the limits it imposes upon us. It is hardly possible today to appreciate the courage it took to rank in the same category, and to subject to the same laws, phenomena so seemingly disparate as falling bodies, the vibration of strings, planetary motions, and collisions.

We must nonetheless add that this quest for universal principles did not begin with Newton, because Descartes had engaged in it before him. The difference between the two is that the French philosopher did not have either the luck or the occasion or the genius to find the true principles of dynamics, and those he did propose were incomplete, when not false. He also probably overestimated the power of his method, based more on reason than on experience. This priority, and the fact that Descartes must be recognized as a great philosopher (while this quality was less manifest in Newton's case) are the reason for the name *Cartesian project*,* given by Husserl and Heidegger. Following the latter, we shall adopt this expression, obviously without sharing his aversion for the project itself.

This project, which has nowadays become almost a fundamentalist doctrine among most scientists, is based on the claim that nature obeys some universal principles that are expressible through logical and mathematical means. If we take a cold look at this idea, we must admit that there is in it an element of madness: how can one presume that the multitude of objects and phenomena in nature, their swarming diversity, the stuff of poetry and fantasy, that all that could be disciplined with an iron hand? It is certainly due to the accumulated weight of so many discoveries, to the evolution of minds caused by history, and to the effect of a systematic indoctrination, that this idea gradually became sufficiently conventional to be embraced by some so intensely that questioning is no longer necessary, and to make of it an article of faith, the stronger because not pronounced.

This ambitious perspective appears from the beginning, in Newton's definition of the framework of dynamics: space and time are absolute. For him, physical space is no longer structured along the horizontal and the vertical, purely terrestrial features, but it is absolute: "Absolute space, in its own nature, without relation to anything external, remains always similar and immovable. Relative space is some movable dimension or measure of the absolute space; which our senses determine by its position to bodies; and which is commonly taken for immovable space." There is similarly an "absolute, true and mathematical time, of itself and of its own nature," which "flows equably without relation to anything external . . . , relative, apparent and common time, is some sensible and external (whether accurate or unequable) measure of duration by the means of motion, which is commonly used instead of true time."

Few passages in physics have been so copiously quoted and commented on, and justifiably so. Everything is there: a claim of absoluteness, almost metaphysical and very close to one of Kant's a priori judgments of reason. But Newton is less categorical, for he indicates an experimental procedure to identify, in principle, this absolute space, and he also hints at the possibility of believing otherwise. What's more—and proof of his superb cleverness—it turns out that reference to the absolute solves beforehand all difficulties resulting from inertial effects (centrifugal force and others), by allowing them to be deduced from the principles, instead of having to be analyzed in themselves. And despite all that, this fertile simplicity is not sufficient to guarantee a definite and unshakable truth, as Einstein was later to demonstrate: "Subtle is the Lord."

The principles proposed by Newton are well known. There are three in all. The first one is none other than Galileo's principle of inertia, in Descartes' version and reformulated within the setting of absolute space and time: a body not subject to any forces moves (in absolute space) in a straight line with constant velocity. Then comes the equality of action and reaction, a principle already known in statics. The third one, often called *the* fundamental principle of dynamics, is an old friend of college students: the product of the mass of a body with its acceleration (in absolute space) equals the total force acting on it. The notion of acceleration, which plays a central role here, is based on another of Newton's great discoveries, differential calculus. If the force acting on a body is known, the third principle can be translated in terms of differential equations, whose solutions then express the position coordinates of the body as a function of time. Newton supplied the meaning of these equations, as well as a solution method: integral calculus, another of his creations.

Newton's first task is to establish the plausibility of his theory. He does it by a method that would show up many times in the development of physics: recovering some results, already known in the form of empirical rules, as logical or mathematical consequences of newly stated principles. The results in question involve the motion of the pendulum, falling bodies, and properties of collisions.

But his greatest triumph is, as we already know, the theory of gravitation—which is essentially no different, after all, than the above results. The question was to retrieve Kepler's empirical laws from the general principles. The only serious difficulty is to determine the exact form of the gravitational force between two bodies—the sun and a planet, for instance. But this can be achieved by exploiting two of Kepler's laws. The law of areas is really the universal signature of a central force, that is, in this case, a force directed along the line joining the planet to the sun. And from the empirical rule relating the major axis of a planet's orbit to its period (or planetary year) it follows that the above force must be inversely proportional to the square of the distance. Now we know everything we need to solve the equations of motion resulting from the fundamental principle of dynamics. Their solution would confirm that the planetary orbits are, as expected, ellipses, with the sun at one focus.

Let us remark, in passing, that a historian of science would surely object to the way we derived the relationship between Newton's principles and Kepler's rules. He or she would argue that the discovery of the famous inverse squares law was much more involved, filled with intuitions and hesitations, not to mention controversies, and that the said law had been anticipated, if not discovered, by Hooke. The historian would also point out that the arguments we presented are in fact fairly recent. And certain philosophers of science would add that we have just confirmed the intrinsic dishonesty of scientists, who misrepresent history to make believe there is a method where, as Feyerabend claims, there is nothing but chance, chaos, and guessing. Such scruples are legitimate, and we owe the reader an explanation. First, we remind him or her that this book is only incidentally concerned with history, and that its unique aspiration is to understand—out-of-timely, we might say, if that would make sense. We thus feel justified in taking advantage of the latest ideas when they are the clearest and the most relevant. As for the scientific method, this question will have to wait, for it is subtler than generally admitted, especially when one wishes to deny its existence.

As we close this parenthesis and go back to the theory of gravitation, let us examine a difficulty that was the source of much discussion. The gravitational force was supposed to act directly between the sun and the planet, in spite of the vacuum separating them. But how is it possible that an action or, as we would say today, information, could be carried by vacuum, by total emptiness? This conceptual difficulty of action "at a distance" defies common sense, a forbidden equation between vacuum's nothingness and a force's being, a philosophical non-sense. Newton knows it well, and he openly admits his embarrassment in the general scholium of the *Principia*'s third edition. Einstein would bring up yet another questionable feature of this force: the fact that it depends only on the distance separating the sun and the planet at the very instant the action takes place. This seems to imply that the sun has some kind of instantaneous perception (or information) of the planet's position.

Instantaneous action at a distance; such is the original flaw, as it were, of the Newtonian theory of gravitation. It was rapidly forgotten by most laborers of science. As a matter of fact, without yet realizing it, they were unconsciously shifting from an intuitive science, where everything can be visualized and is in agreement with common sense, toward a science involving formal elements that are essentially unintelligible.

This shift would be more intense at the turn of the eighteenth and the nineteenth centuries, with the purely mathematical works of Laplace, Lagrange, and, shortly afterward, Hamilton. Lagrange and Hamilton, in particular, render Newton's principles in a mathematical form quite different from their original formulation. The notion of action is central here. Unlike the concepts employed by Newton, this notion is purely mathematical, without any intuitive, visual, or analogical content. The action is an integral over time involving the difference between the kinetic and the potential energies. We can surely make sense of their sum—it is the total energy—but their difference? What's more, action means nothing by itself, it is only an intermediary: actual motion has a kind of magical property, which is to minimize the action (the *principle of minimal action**). Why a minimum, or even a maximum? We can only wonder, without expecting to understand, without "seeing" anything, because we do not know what "action" is or where it comes from.

Thanks to the methods of Lagrange and Hamilton, it was possible to go straight to the heart of the computations in dynamics, and often perform them much more efficiently. But these purely mathematical qualities have nothing to do with the essence of physics. More efficient calculations do not entail higher conceptual content, so no one could claim that the arrival of the new methods called into question the foundations of science. And yet. . . .

35

Waves in the Ether

It is certainly not our intention to go into the details of the history of science; we must nevertheless say a few words concerning optics. There are several reasons for this. First of all, unlike dynamics, which deals with the motion of concrete objects, of entities having no secrets, optics poses a big question: What is light? In this respect, optics demands a level of comprehension that is deeper, more difficult to satisfy, than that of dynamics. This branch of physics is also one of the most outstanding examples of the coherence of science—which will be one of our leading threads—as shown by its eventual unification with electromagnetism. It is worth examining it, even if briefly.

Antiquity had bequeathed us the appealing hypothesis, coming from the atomist school of Leucippus and Democritus, according to which light is made of a particular kind of atoms. These are emitted by luminous bodies and, after bouncing off an illuminated object, are captured by the eye. This hypothesis, put forward in Lucretius' *De natura rerum*, is also a good example of what the ancients considered an explanation to be: a satisfactory image that a person composes for himself or herself, and which can be communicated to others orally.

Scientific optics, that is, the search for empirical rules through experimentation, really begins with Descartes' *Dioptrique*, published in 1637. Some of these rules come from the distant past. The linear propagation of light or the laws of reflection on mirrors, for instance, were known to Archimedes. Other laws are new, such as those concerning refraction (discovered earlier by Snell); they govern the change of direction of a ray of light as it crosses the surface separating two transparent media (water and air, say). Descartes also draws a number of consequences. He impresses his contemporaries with his theory of rainbows, which he explains in terms of reflections and refractions in droplets of water.

This is a remarkable explanation, being a case where science unraveled the secret of one of nature's most intriguing mysteries. Rainbows are not the stuff philosophers usually ponder over; nor are they one of those deep enigmas that strain the reflective mind. On the other hand, they delight the poet, who finds in them a multitude of interpretations. When science reveals the true nature of such a lyrical phenomenon, it exposes itself to be taken by some for an enemy of poetry and dreams. Others, on the contrary, learn to see better, to make out the faint secondary rainbows predicted by the theory and to marvel at their existence. Poetry has not been banished; it only assumes a new dimension, that of the order filling the universe. Whatever the case, by the multiplication of similar discoveries science in the end transforms our imaginary world as well.

Those with a desire to understand are not less imaginative than the dreamers. The birth of the wave theory of light provides a good illustration of this fact. It is a highly revealing episode, for it exposes a facet of science that is crucial if we are to grasp its real method: when reason is most confident in its own power, it may deceive itself and err, to arrive at the right idea only by a lucky strike.

Method is precisely Descartes' main concern. He advocates one that is based primarily on reason, much more than on experience. It consists in decomposing each problem into smaller and simpler ones, until the solution imposes itself as evident. Its ultimate goal remains a complete synthesis, by which the mind should be able to apprehend everything in a clear and thorough fashion. This approach has made Descartes one of the most typical exponents of *realism**—the belief in the possibility of achieving a perfect knowledge of reality. An essential component of his vision of physics is the identification of matter with extension or, we may say, with space. Space is also matter, and even its most intimate parts can therefore be in motion. As for light, he rejects the corpuscular hypothesis—because it is unsuitable for explaining the changes in direction taking place in refraction—in favor of a waving of matter (or of extension) similar to sound waves, whose character was by then understood.

This wave hypothesis will gain ground during the next generation, thanks to Robert Hooke and Christian Huygens. Light is still supposed to propagate through a transparent medium, as sound waves do through air. Each individual point of the medium vibrates, and all these vibrations together, as they propagate from point to point, form the wave. Huygens works out the details of this theory and translates it into precise mathematical language. From this he infers that light waves must travel in straight lines, and he also succeeds in accounting for the laws of refraction.

*Diffraction** phenomena, observed long ago by Leonardo da Vinci and later studied by Grimaldi, also confirm the wave idea. One of their manifestations is the lack of sharpness of a shadow, its slightly fuzzy edge when examined closely; or the way microscopic particles or a spider's web diffuse light. All this suggests that light wavelengths are not inaccessible, even if they are very short.

The wave assumption would, however, carry with it a deficiency from birth that would persist long after Descartes' initial ideas had been abandoned: If something does vibrate, what is it? When light propagates through some material medium there is no problem, atoms can do the trick. But what happens in vacuum? Huygens' reply—the *ether** hypothesis—will mark the history of physics for years to come. For Huygens, the ether is an immaterial, all-pervasive medium that penetrates matter and is everywhere present, even where there appears to be nothing but emptiness. It is nonetheless a mechanical medium, because it vibrates when traversed by light waves.

If we agree to define classical science as a description of reality using concepts that the mind can readily interpret, then the introduction of the ether by Huygens, shortly before Newton had recourse to action at a distance, signals a crack in this conception. But let us not jump the gun.

The wave theory of light had to compete with the particle hypothesis, the latter helped by the weight of Newton's endorsement. We shall not be concerned with the details of this great controversy, which the discovery of interference was going to put to rest for a long time. Thomas Young was the first to observe *interference** effects, in 1801. To perform his classical two-slit experiment, he illuminated one side of a sheet containing a pair of narrow parallel slits and observed alternating dark and light fringes on a screen placed on the other side. Augustin Fresnel improves the procedure by employing two slightly slanted mirrors instead of slits. This allows for better observations and a more systematic experimental study. Waves on the surface of water also result in interference effects. It is then natural to interpret the outcome of these experiments as supporting the wave hypothesis, and this is precisely what Young and Fresnel do. At the same time, Fresnel further develops Huygens' theory, which allows him to account for the interference fringes in a quantitative way, and to apply the theory to diffraction phenomena.

At this point took place an incident already reported one thousand times, but which we cannot avoid telling once again. It will give us the occasion to show an example of what is called a crucial experiment. In 1819, the Paris Academy of Sciences had charged a jury with the evaluation of works on diffraction phenomena. The members of the jury were Biot, Arago, Laplace, Gay-Lussac, and Poisson. Fresnel submits for the occasion a detailed account of his ideas on the question, complemented by additional calculations. One cannot but admire the jury's commitment to what will turn out to be a thorough investigation of the problem. Poisson, the fearless mathematician who used to say, tongue in cheek, that there exist only two things making life worth living, doing mathematics and teaching it, undertakes an exhaustive analysis of Fresnel's work. Using sophisticated mathematical tools, probably too advanced for Fresnel, Poisson calculates the intensity of light far inside the shadow of a circular screen. The result of his calculations defies comprehension: at the very center of the shadow there should appear a bright area of intensity equal to that of full light. This conclusion seems an aberration. Poisson is then on the verge of rejecting Fresnel's entire work when he decides to share his puzzlement with Arago. The latter points out that the bright area in question, being quite small, might easily have gone unnoticed until then. He then performs the experiment only to verify, to his great astonishment, that the incredible bright point is actually there!

This type of incident is usually more convincing than a whole collection of patiently gathered pieces of evidence. After the Poisson–Arago demonstration, the wave hypothesis seemed all but inevitable, and no one doubted any more that light was a vibration. Let us conclude the story here, at this blissful moment when only one cloud remains—that of the disturbing ether.

The Beginning of Electromagnetism

For a long time, it was believed that science could advance by what was called the induction method.[1] The basic idea is that an attentive examination of the facts should permit identification of the

[1] The problem of induction, as it is normally defined in the philosophical literature, consists in correctly estimating the plausibility of a scientific rule or principle, knowing that we can have access to only a finite number of examples of its

appropriate concepts, and even suggest the rules or laws they are subject to. The history of the origins of electricity and magnetism provides a good example of this conception. At the same time, if we ignore some premature attempts to introduce a theoretical basis, it demonstrates how a purely experimental approach can go on for several generations.

Antiquity did not know a great deal on the subject, little beyond the fact that rubbing an amber (*elektron*) stick provoked some strange effects, and the mysterious properties of an iron oxide such as magnetite, which the Chinese would exploit to built the compass. "Subtle is the Lord," because we now know that under those humble appearances lurked the most important forces at the heart of matter.

We must wait until 1729 to discover, thanks to Gray, that an electrically charged object may electrify another, and that there exist conductors and insulators. In 1730, du Fay observes attractions and repulsions among electrified bodies. This fact leads him to assume the existence of two different types of electricity, one positive and the other negative; objects with electricity of the same type repel each other. These forces would permit quantification of the still vague notion of electricity using the concept of electric charge: if A, B, and C are electrified bodies, then the charge of A is said to be equal to the charge of B (or twice as large) if, after placing A and B in turn at the same distance from C, they are subject to a force that is identical (or twice as large).

Watson and Franklin show, in 1747, that when two initially neutral bodies become charged by mutual interaction (by rubbing them together, for instance), their charges are of equal magnitude but of opposite signs. They conclude that the charges are not created during the interaction but were already present in matter and canceled each other out exactly: electrification separates the existing charges permanently.

Priestley, Cavendish, and, especially, Coulomb (whose main contributions appear in 1785) translate the above properties into quantitative terms. Coulomb designs an extremely sensitive torsion balance capable of measuring very small forces. The force between two minute electrically charged bodies turns out to be very

---

application. The method of induction discussed here has a much more limited scope.

similar to Newton's gravitatonal force (inversely proportional to the square of the distance). This force is also proportional to the product of the charges, while gravitational force is proportional to the product of the masses. Such similarities will allow the theory of electrostatics to make some quick progress, by transposing to electricity the results of Laplace, Poisson, and Gauss on gravitation. As a consequence of this approach, the notion of electric potential is soon introduced. The invention of electric batteries (Volta's dates back to around 1800) makes it possible to generate electric currents routinely. In 1826, Ohm establishes the empirical rule relating current and potential difference, leading to the notion of resistance.

Strong analogies between electricity and magnetism had been noticed all along, but without any concrete consequences. The connection between the two, and their belonging to a common and larger setting, is suggested in 1820 by a discovery due to Oersted: a current-carrying wire exerts a force on a magnet. Soon afterward, Biot and Savart state the quantitative rules governing the link between electric currents and magnets, and Ampère gives them a simpler form, in which each element of the current acts separately. Still in 1820, Ampère measures the force between two electric currents and discovers that a current-carrying coil of wire behaves like a magnet.

Thus, an electric current could create magnetic effects and also behave like a magnet. This immediately suggested the possibility of the inverse phenomenon: Could a magnet produce electricity? It was only a decade later, in 1831, that this question was answered in the affirmative by Faraday—the hitch was that it takes a moving magnet to "induce" an electric current in a nearby wire. Faraday also invented the condenser, and noticed that the presence of insulators significantly affects the intensity of electrical forces. The study of those so-called dielectric media would prove to be very important.

A Turning Point: Maxwell's Equations

One might have thought, around 1840, that the totality of electric and magnetic phenomena was essentially known, but a closer look showed that something was still missing. The present knowledge includes laws expressing how charges, currents, and magnetic moments generate forces that act on other charges, currents, or magnetic moments. There was also an assortment of other laws, such as Ohm's law or the law of induction. However, a review of all these empirical rules reveals that, unlike Newton's principles, they do not constitute a comprehensive dynamics. In other words, they do not allow one to infer the future evolution of charges and currents from their values at a given instant, nor do they explain how the electric and magnetic properties of matter are determined.

But this mathematical deficiency was not the main concern of certain researchers. Their dissatisfaction stemmed rather from an unfulfilled desire to understand: they lacked a satisfactory image of the motion of the charges and magnetic particles present in matter, and whose interactions might have produced the observed empirical phenomena. If understanding meant being able to "see" what things are, they definitely did not understand. The great adventure about to begin would be precisely to make sense of the wealth of experimental knowledge that had accumulated for many generations. Its goal was an almost impossible dream: to achieve full coherence, though at the same time it would mean the end of many intuitive representations.

The best mathematicians among the physicists—Gauss, Ampère, Biot, Savart, and others—had always concentrated their analysis on certain notions—charges, currents, magnetic dipoles—that are relatively concrete and easy to envisage. Even if their theories occasionally referred to electric or magnetic potentials, these were construed merely as mathematical devices that facilitated the computation of forces, just as in the theory of gravitation.

But that was not Faraday's point of view. When he saw iron filings orient themselves under the influence of a magnet, he expected something real to be behind those lines of the magnetic "field," something more important and significant than the forces acting at a distance postulated by the theoreticians. He rejected the very idea of action at a distance, and wanted to understand how the filings or the molecules of dielectrics form patterns and progressively influence one another. He then set out to build models of the behavior of matter where two fields, an electric and a magnetic one, played the central role. Let us recall, using the example of the iron filings, that a field is "something" having a magnitude and a direction (a vector), defined for all points in space, and possibly changing with time.

Faraday was a first-class experimenter and a physicist of genius, but he was also self-taught, and lacked the necessary knowledge to put his ideas in mathematical form. And so his models, even if most ingenious, fell short in regard to their quantitative aspect. The honor of fleshing them out would fall on his disciple James Clerk Maxwell (1831–1879).

To better understand Maxwell's contribution, let us recall the principal laws of electromagnetism and the form under which they were known at the time. There were—and still are—four of them. The first one is Coulomb's law, which gives the force between two charges. Thanks to Gauss, it was also known how to express this force using the notion of electric field. The second law gives in a similar fashion the pair of forces between two elements of a magnet. Using a magnetic field, this law can be put in a form closely resembling the first one. The third was Ampère's law (or Biot and Savart's). From Faraday's point of view, it expresses the value of the magnetic field generated by a current. Finally, the fourth one is the law of induction discovered by Faraday himself; it gives the magnitude of the electric field induced in a circuit by a variation of the magnetic flux through the circuit. One might say that, thanks especially to Faraday, the laws of electromagnetism could be formulated using either fields or the forces acting between charges and currents.

Maxwell's first task, in 1855, will be to straighten up all that. He begins by examining the laws that were already known. It is an arduous enterprise, since the mathematical tools of the time were ill-suited for the task. The modern methods of vector calculus not being available, Maxwell has to combine mathematical techniques with physical intuition, and does not hesitate to use hydrodynamic analogies. One of the most remarkable consequences of his results will be the possibility of rewriting electrical energy as a function of the electric field alone, without any reference to the distribution of charges.

In 1861–1862 Maxwell goes farther: he tries to "understand," that is, to correctly decipher reality. He wants to "see" what goes on inside matter under the effect of a field, and cannot help wondering what the ether looks like from the point of view of electromagnetism. It was necessary to assume the existence of the ether, because if an electric force can act through vacuum, this vacuum must be "something" that transmits the effect of the force. Starting with matter, where he "sees" the behavior of the molecules, and then proceeding to vacuum, he is led to imagining a model of the ether that is absolutely fantastic. The ether contains cells, animated by microscopic currents that account for the transmission of the magnetic field. All through the ether, there is a compact network of elastic strings, the lines of electric field through which charges travel. This whole dreamlike construction, which could have been inspired by Jérôme Bosch, is progressively built, exploiting the analogy between matter and the ether. Maxwell analyzes it in the most serious and thorough fashion. He takes into account the forces being exerted among the various components of the system and their effect on motion, in an orthodox manner and according to Newton's principles.

Maxwell succeeds in this way in recovering the familiar laws of electromagnetism, but with an important difference. In fact, only the third law, going back to Laplace and Biot and Savard, needs to be revised. In its previous form, it described how a current generates a magnetic field. Maxwell realizes that it is also necessary to assume that a magnetic field can be created by a changing electric field. This comes very close to the induction law, where a change in the magnetic field can generate an electric field. Maxwell calls the electric source of the magnetic field "displacement current" (even though no motion of charge is involved). Its effect, which appeared to be very weak under the experimental conditions of the time, has important implications from a conceptual point of view, for it permits one to obtain the "electromagnetic" field equations—the celebrated *Maxwell's equations.** These have two chief properties: they provide a field dynamics, in the sense that they can be solved for any future time if the values at some initial time are known; and they guarantee the conservation of energy, if one identifies correctly the energy due to the magnetic field.

As a result of all this effort, Maxwell finds himself in an unprecedented situation. On the one hand, he has obtained new physical laws, consistent with the empirical rules previously discovered. What's more, his new laws exhibit a superior mathematical and physical coherence. But on the other hand, the model that led him to these results, with its ether packed with cells and strings, is highly improbable, even for its creator. The final product was good, but the mold from which it emerged would have to be destroyed.

And so, in 1864, Maxwell returns to the subject, this time using an altogether different method. Modeling the ether is out of the question; he starts, on the contrary, from concepts stripped of physical meaning to the point of becoming almost purely mathematical. The "things" that can be everywhere are the electric and magnetic fields; and wherever they are, there is energy. The electrical component of this energy is identified with a potential energy, while the magnetic component is considered as kinetic energy. From a mathematical point of view, the method is quite different. Having identified the variables (the fields) and the two forms of energy, it is possible to apply the abstract dynamical methods of Lagrange and Hamilton and their minimal action principle, without further knowledge of the nature of the system. This is basically what Maxwell does, and also what Hertz will do shortly afterward, when he will refine the method. And so, almost in an automatic fashion, Maxwell obtains the equations of dynamics for his system of fields. These are none other than the same equations he had derived earlier by entirely different methods.

This last accomplishment of Maxwell marks the turning point in the transformation of physics. The curtain falls on classical physics, if we understand by "classical" an explanatory physics, where reality is visually represented in a way that can be fully grasped by intuition. This classical physics has just been replaced, for the first time in broad daylight, by a formal physics whose basic concepts (the fields, in this case) have a strong mathematical flavor and, especially, whose principles (Maxwell's equations, or the mathematical equivalent of Lagrange's principle of minimal action) have become purely formal and mathematical, a kind of abstract and rather obscure essence of Newton's first principles. By the same token, Lagrange's principle of minimal action took on a new significance, and became in a certain sense *the* leading principle of dynamics.[2]

---

[2] The principle of minimal action, in its classical form, is still surrounded by a certain mystery. But this mystery is none other than the mystery of quantum mechanics. Richard Feynman derived this principle from quantum mechanics in 1946.

When someone asked Hertz which principle lay at the basis of Maxwell's equations, he would reply, the equations themselves. Richard Feynman, one of the most intuitive physicists of our time, also used to tell his students that it is impossible to imagine the electromagnetic field. In fact, after Maxwell, physics is really no longer something one can visualize with the imagination and communicate in ordinary language. Its concepts cannot be completely rendered without at least the help of mathematical language. The latter has presently become an intrinsic component of physics, and not only of the quantitative form of physical laws. If Voltaire was able to explain Newton, no philosopher, however persuasive, will ever explain Maxwell to a delightful marquise.

And yet, the fruit was there, since thanks to Maxwell's equations one could verify, as Hertz would do in 1888, that the electromagnetic field can vibrate—a vibration that is also light.

In these final years of the nineteenth century there are other foretelling signs of intuition gradually going blind. And little by little, formal concepts come to the rescue; thus, entropy replaces heat. At the very moment some were ready to declare the edifice of physics almost complete, the time was ripe for another physics to be born.

Classical Mathematics

I<small>T IS UNTHINKABLE</small> nowadays that a philosophy of knowledge should dispense with a serious reflection on mathematics. To stop at a meditation on logic, as certain authors do, is unacceptable for a modern science where mathematics, in all its profusion and sophistication, pervades the articulation of concepts and laws.

The real difficulty, and the reason why many avoid mathematics, consists in giving it its proper place without spending long years of study. Its vastness is truly impressive, and, like the ocean, it contains plenty of delicious foodstuff. Some people, the most vulnerable to its seduction, plunge into it for the rest of their lives; others are happy with swimming from time to time near the beach; yet others, disdainful cats, refuse to even dip a foot in it. And so mathematics, with its uncharted borders, may appear welcoming or hostile. It is nevertheless unavoidable, and Plato's adage, "No one may enter here who is not a geometer," has never better indicated the passage to philosophy.

We shall limit ourselves to the essential, to developing enough on which to base the theory of knowledge. We shall also show, with the help of history, that the crude formalism, arrogant at times, that characterizes mathematics, is the result of necessity, of consistency, and not of a deliberate esotericism. Everything else we shall omit. The principal drawback of this approach is to leave out the ideas and methods that account for the richness, the fertility, and the vision of mathematics. It will also be necessary to exclude most of the concepts and computation methods so essential to other sciences that this is precisely the reason we are forced to make this incursion. Since we cannot help it, we shall leave it at that.

CLASSICAL MATHEMATICS

Since when have humans been fascinated by numbers and figures? From time immemorial, practically all ancient civilizations have believed that certain numbers have a sacred character, varying with time and place. Is it possible to imagine eleven muses, seventeen gods of the Olympus, or eight days for the creation of the world? The fact that small numbers should fascinate us more than others is understandable, but why ought three, four, seven, and twelve be more important than five or nine—themselves superior to six and ten—while eight and eleven do not mean anything to anyone? Beyond them, all are very large numbers.

The attraction of certain geometric figures, such as the circle, the equilateral triangle, or the square, may be explained by their multiple symmetries. But how could the peculiar claim that the circle is the only perfect curve so mark the audacious minds of the Greeks, that they rejected all the others as unworthy of celestial bodies? There has always been a feeling of perfection, of divine, attached to numbers and figures; a strange inclination, sometimes present in children, which seems to suggest that its structure exists in our brain.

It is generally believed that mathematics was born out of practical experiences: we can trace out a circle using a string; the right angle between the vertical and the horizontal ensures stability; a rectangular form guarantees a constant area for a field swept by the mud of the Nile; and to form the necessary right angles one might construct a right triangle with a rope in which knots are separated by distances in the progression 3, 4, and 5. Thales discovered very early that the parallel rays of the sun, visible when they pierce a cloudy sky, could be used to measure the height of a tree by comparing its shadow to that of a stick. From all this follows both an interest in fractions and the existence of a close correspondence between figures and numbers.

Pythagoras went farther, with his famous theorem about right triangles, where the above correspondence manifests itself clearly. He might have guessed it from a simple drawing, but it is certain that mathematics already possessed its basic logical tool when the unknown Pythagorean proved that no fraction can measure the diagonal of the square. Logic is mathematics' twin sibling; and only logic makes *proving* possible. But we have already said this.

The discovery of the irrationality of the diagonal reminds us of Thomas Kuhn's famous theory, according to which the progress of science proceeds by *paradigms*,* that is, by examples that are so striking and inspiring that they command an almost religious acceptance. The discovery of the irrationals was a kind of paradigm of paradigms, because it contained the seed of an infinite science. Our remarkable Pythagorean must have lived shortly before Socrates. Plato already knew some nice mathematical results, and his contemporary Eudoxus had discovered a great number of theorems in geometry and in the theory of numbers. Mathematics will rapidly reach maturity with Euclid of Alexandria, of whom we know only that he lived several years after some of Plato's disciples (the master having passed away in 347 before the Christian era) and before Archimedes (287–212 B.C.).

Even if history has retained only a few illustrious names from this era, Euclid's mathematical style betrays without the slightest doubt the long discussions—of the passionate kind Greeks could have—that had preceded his work. In it we find a pursuit of the simplest hypothesis, an effort to provide only irrefutable arguments, and an elimination of the unnecessary that can only be the result of endless revisions, brought about by endless objections in a race toward perfection. Plato's early dialogues remind us of such animated talks and are a reflection of these controversies. Some of these are perfectly legitimate, such as the one concerning Euclid's famous postulate of the parallels: there is only one line parallel to a given line and passing through a point not in the latter (by "parallel," he meant lines that never meet). Remember that Pythagoras believed that the stars were fixed to a celestial sphere, and that others thought that there was no space beyond it; then imagine the kind of debate that the presumed existence of parallels might have triggered. Euclid's postulate assumes in effect an infinite space; therefore, for the spirit of the time, it conceals a hidden cosmogonical hypothesis. One might be tempted to drop this postulate, but then many precious results could no longer be demonstrated—that the sum of the angles of a triangle equals a flat angle, for instance.

The existence of such dilemmas explains Euclid's care in clearly distinguishing the different kinds of assumptions: axioms, postulates, definitions and hypotheses. An axiom is an immediate truth that no Greek would have questioned during a discussion, for instance, "Two distinct (straight) lines that meet have a unique point in common." A postulate is a statement that we assume to be true, even if its status might have raised questions in the past. The truth of a postulate is actually taken for granted by those playing the mathematical game, because they know well that if they denied it, the game would lose part of its charm. Only later would this cautious distinction between axioms and postulates be abandoned, and the latter taken to be indubitably true, just like the former. As for Euclid's definitions, they are of many sorts. Some are perfectly clear and are true definitions, in the sense that they allow us to argue without ambiguity—the definition of a circle, for example: all points at a fixed distance from another point called the center. Others are more like words uttered without much conviction, expressing something like "I do not know how to put it, but you no doubt see what I mean"—the straight line is "defined" as that which lies evenly on all its points. Finally, hypotheses serve only to make precise the subject under discussion, and in which context. They usually begin with "let": "Let a triangle have an obtuse angle. . . ."

By the end of antiquity, the accumulated knowledge in geometry is considerable, from properties of triangles, polygons, circles, and conics (ellipses, hyperbolas, and parabolas) to those of other curves generated by simple motions. In space, it concerns principally polyhedra, spheres, cones, cylinders, and ellipsoids of revolution. We must not forget trigonometry, plane and spherical, extremely relevant in organizing astronomical observations.

Regarding arithmetic, a science as useful as it is arid, we shall say nothing besides remarking that it would give birth to algebra, thanks to Diophantus, who lived in Alexandria in the third century after Christ. We know little about him, except that he spent one-sixth of his life as a child and one-twelfth of it as an adolescent; that he lived seven more years before having a son who lived half as long as his father, and, finally, that Diophantus outlived his son for a period equal to one-sixth of his own life. The result of all these calculations would put Diophantus' age at eighty-four years. Algebra might have been invented by a teacher. For it is easy to imagine that the poor individual, tired of repeating arguments that led to the same arithmetic calculations, finally realized that the particular values of the numbers are irrelevant, and that all that matters is the pattern of the operations on these numbers. Be that as it may, no one considered it necessary to axiomatize algebra à la Euclid, since this had already been done for the theory of numbers, and besides, algebra—it was believed—was nothing more than a collection of convenient "recipes" summarizing some known arithmetical processes.

The development of algebra was long delayed by the cumbersome numerical notation used in the Greek and Roman world. This obstacle was later removed by the Arabic civilization, which employed the "arabic" numerals we still use today, as well as by the introduction of zero, a fabulous idea imported from India. The invention of the negative numbers followed. Algebraic notation proper, that is, the symbolic expressions representing operations on numbers (using signs, such as our +, −, =), was also making progress.

The importance of notation in mathematics cannot be overemphasized. A well-chosen notation suggests the right operations and liberates the mind from pointless distractions, while an ill-chosen symbolism may be a hindrance to reasoning. From the point of view of logic and rigor, notation is irrelevant, but it has a bearing on the relationship between imagination and formalism. An efficient notation should be suggestive, meaningful, suited to our imagination as much as to the subject matter. Is this the reason why algebra would be a product of the Arabic civilization, which rejects any explicit images and gave algebra its name?

Antiquity knew how to solve only quadratic (or second-degree) equations and systems of linear (first-degree) equations. When mathematics resumes in Europe, during the Renaissance, Cardan and Tartaglia discover a method to solve equations of the third and fourth degrees. In so doing, they bump for the first time into the imaginary numbers, whose prototype is the square root of −1. There are cases where while solving a cubic (or third-degree) equation—to obtain a definite numerical value for the unknown—one needs to introduce in the solution process imaginary numbers, which act as a kind of intermediaries (but do not appear in the original equation or in its final solution). This strange phenomenon emphasized, for the first time, the singular nature of algebra as compared to the well-codified mathematics of Euclid, thus making it difficult to view algebra as a mere appendix of arithmetic.

In the seventeenth century, it was geometry's turn to be refurbished by the invention of analytic geometry, due to Descartes and Fermat. The basic idea is to define a geometric point by its coordinates—numbers that locate the point relative to a system of axes. A plane curve is then completely characterized by the equation satisfied by the coordinates of its points. In this way, many problems in geometry can be reduced to algebraic calculations.

It was a considerable step forward, for geometry could now get out of the straitjacket imposed by Euclid's methods, where the only curves were those that could be obtained out of planes, lines, circles, spheres, and cones. The new techniques also permitted an easier manipulation of certain curves introduced at the end of antiquity—for example, those arising as trajectories of motions, such as the famous cycloid traced out by a nail fixed to the edge of a rolling wheel. These novel methods also raised some subtle difficulties which could surreptitiously modify the nature of mathematics. The huge portions that had just been added to Euclid's geometry were based solely on algebra, and no longer on axioms of a geometric character; but, as we have seen, algebra suffered from some logical deficiencies. How to resolve these difficulties? The kind of answer given in practice to these misgivings was more reminiscent of Alexander cutting the Gordian knot than of Euclid's prudent logic; it was the answer of conquering armies: "Ignore the obstacles and move forward."

No time should indeed be wasted, for there is too much loot to be taken. The seventeenth century ends with a glorious deed: the invention of integral calculus almost simultaneously by Newton and Leibniz. It was, in fact, the culmination of a constant progression to which all the great mathematicians of the time had contributed to some degree. But it is also literally a torrent that begins, a prodigal flow, carrying along as many new problems as revealing solutions; new and fascinating results, enough to fill Euler's twenty-three thick volumes and still leave plenty for other mathematical giants: the Bernoulli brothers, Lagrange, d'Alembert, Laplace, and Fourier, who continue the discoveries into the beginning of the nineteenth century.

How distant was then the Greek ideal, when Euler did not hesitate to write that the sum of $1 - 1 + 1 - 1 + 1 - 1 + \cdots$ is ½, even if the successive partial sums are only 1 and 0. And yet, more often than not, these unceremonious methods succeeded beyond all reasonable expectations. When the thrust began to die out and it was possible to step back and take stock, many began to wonder how to render mathematics productive again, and also how to reconcile it with logic's demands for certainty. The story of the answer to the second question will occupy the rest of the chapter. As we shall see, this second answer contained the key to the first one as well.

RIGOR AND PROFUSION IN THE NINETEENTH CENTURY

A mathematician is no more naturally inclined to rigor than a politician; both only put up with it when it becomes inevitable. Having seen how rigor came upon the Greeks, we shall now examine how it will impose itself again on their successors, to lead them to a perfection of formalism. Two partly opposite movements take place during the nineteenth century, a blessed time for mathematics. One of these favors an increase in rigor, while the other continues, after a short pause, the flow of discoveries.

The movement begins with Carl Friedrich Gauss (1777–1855). He was called the prince of mathematics, a title that expresses, just as that of prince of poets, his peers' admiration for a life's work that is as sound as it is prolific. In any case, he deserves recognition for having demonstrated that rigor is the mother of invention.

Before Gauss, it was generally assumed that any given algebraic equation must have roots, which can sometimes be complex (imaginary) numbers. D'Alembert had tried, unsuccessfully, to prove this fundamental fact upon which large parts of algebra and analytic geometry rest. Laplace came closer, but it was Gauss who around 1815 finally provided a convincing proof. The use of complex numbers in algebra had been definitely mastered.

Gauss possessed an acute sense of logical rigor, stronger than that of Euclid, and anticipatory of the future evolution of mathematics. His personal notes show him clearly ahead of his time, and he did not hesitate to question, in private, the postulates of Euclidean geometry. He was particularly troubled by the postulate of the parallels. Others before him had tried to deduce it from the remaining postulates, and so had obtained, unwittingly, the first fragments of non-Euclidean geometry. Gauss' approach is completely different, because he has real doubts concerning the truth of the fifth postulate. Besides, he is not at all convinced of the necessary harmony that should exist, according to Kant, between the space conceived by the mathematicians and the space perceived by our senses. His work on geodesy offers him the occasion to verify whether the sum of the angles of a large (physical) triangle, having mountain summits as vertices, is really equal to two right angles. This is precisely what he obtains, but the inevitable experimental errors still leave room for a reasonable doubt. However, he knows better than to voice his scruples in public, since Kantianism had by then become the established philosophy in Germany, and Gauss hated above all getting involved in sterile debates. So he decides to keep quiet.

Others, less cautious than him, are going to venture onward. Lobatchevski and Bolyai go first. They develop, around 1830, a geometry where the number of parallels to a given line, and passing through the same point, is infinitely large. Riemann follows, by showing in 1854 that there are other possibilities, such as geometries without any parallels to a given line. The predictable stormy controversies ensue in scientific circles, but eventually everyone would come to realize that the logical consistency of the new geometries cannot be denied. What's more, it is possible to construct models of some of these geometries inside a Euclidean space. For example, one of these models is provided by an ordinary sphere if by "line" we understand great circle. Such discussions would result in a better understanding of the fuzzy and conventional aspects of the old definitions, which used to be interpreted with a strong dose of visual intuition. As for the results obtained, they reveal a new phenomenon, one that will recur often in the future: the quest for greater rigor is not an idle exercise in repetition, but can serve to expose new, hitherto ignored possibilities.

During those same years, rigor also enters the foundations of analysis. The notion of the integral still hesitated between an intuitive formulation—using areas, volumes, or masses—and another one based on the idea of the primitive (a function whose derivative is known) which was only valid in some special cases; there were yet others, almost metaphysical, involving the existence of "indivisible" quantities. This confusion was dispelled by Augustin-Louis Cauchy (1789–1857) and Bernhard Riemann (1826–1866). They showed how the integral of a given function could be defined as the limit of a certain sum of elements which increase in number and decrease in size to become "infinitesimally" small.

The idea goes back to Leibniz, the special symbol for the integral, an elongated S, suggesting a generalized sum. In any case, after Cauchy's preliminary results, completed by those of Riemann in 1854, it was clear that this idea of a sum was not just an intuitive approximation but a legitimate definition. The limit in question exists and is unique, that is, independent of, say, the myriad of ways an area may be divided into small pieces. The foundations of analysis seemed then well established, now that the notions of integral and derivative had been worked out (the derivative of a function was also viewed as a limit).

But there was a concept, even more basic than derivative or integral, which had remained totally vague. It was the notion of function—the very objects analysis is supposed to be about. The eighteenth century had been marked by the birth of analysis as a tool in geometry and dynamics. The applications to these fields had been so dominant that no one doubted that the functions arising in analysis could be something else than combinations of polynomials, sines, cosines, exponentials, and other familiar functions.

Soon, new ways of generating functions became available, and that naive assumption began to crumble. The functions of complex variables and Fourier series (to which we shall return) revealed that the former setting was too narrow. The need to break away from it was accelerated by that impulse mathematicians feel to push to the extreme the generality of their results. And so, as soon as the question of which functions were "legitimate" was posed, answering it was considered a priority.

To these calls for serious reflection coming from inside mathematics we must add an external cause of a social character: the level of studies had been rising in universities and other institutions of higher learning. Teaching is a risky and challenging business, often taking place at the frontiers of knowledge, with a potential for error or contradiction. This is how Cauchy's results on integration first came out in his classes at the Ecole Polytechnique. In the meantime, an important change had taken place regarding the social status of mathematicians. Almost all of them now teach, causing the mathematical community to favor a return to the foundations. It is finally the pressure of a confrontation with youth—the history of the Greek origins all over again—that prompts a fresh look at the basis. At the same time, the exercise should prepare teachers to better stand up to objections, and give them an edge in dialectical duels. But there might be another, more noble reason: the well-known but always amazing phenomenon of the greatest virtuosos being also the most attentive in the mastering of the simplest techniques.

The master of rigor and the one who set mathematical thinking firmly "into concrete" was Karl Weierstrass (1815–1897). He is the chief artisan of the clarification of some widely used notions, such as that of a continuous function, or the various types of convergence of sequences. We shall leave out the details, not wishing our account to become too technical. Thanks to Weierstrass, and to some others, analysis finally rests on a solid basis and it becomes possible to verify the scope of its theorems. The more general the theorems, the more fascinating they are; and the more satisfaction they bring to that aesthetic sense characteristic of the mathematical fraternity. This desire for generality constantly calls for sounder foundations, and for greater freedom as well. In this way continues, in a relentless and coordinated fashion, the double exploration of the foundations and of the possible additions to the structure.

Meanwhile, there are further developments in analysis. These result from the study of the systems of differential equations arising in mechanics, in physics, and in applications to geometry—and, of course, from the generalization of these new results. The small family of familiar functions must make room for an army of newcomers: elliptic and hypergeometric functions, and the functions of Bessel, Hermite, Legendre, Jacobi, and so forth—their list is a *Who's Who* of the mathematicians of the time—each of them presenting a particular interest. The functions of one complex variable, initially almost a curiosity, turn out to be essential, and surprisingly useful for all kinds of calculations. Finally, the fundamental principles of analysis raise some important questions that deserve to be discussed separately.

It is not easy to convey the extent of the amazing expansion of algebra and geometry during the same period, because there is innovation everywhere. An old question in algebra, motivated by simple curiosity, will open an unexpected breach: Is it possible to solve—at least in principle—any given algebraic equation by an explicit formula, as is the case up to the fourth degree? The answer is in the negative, and it will be found by Abel and Galois. But the tool employed by the latter, the theory of groups, is much more interesting that the problem it helped to solve. This answer marks one more step toward a formal interpretation of mathematical objects, because equations possess numerical solutions that, even if we know they exist, cannot be exactly calculated.

Systems of linear equations are also studied. They had been known since antiquity, and solution methods had been available for a long time. When the number of unknowns does not exceed three, each linear equation represents a plane in three-dimensional space, and large portions of geometry may thus be put in an algebraic setting. The desire for generality will provoke an interplay between algebra and geometry, a kind of ballet in which each discipline will lift up the other. Geometric thought need no longer be restricted to a three-dimensional space, since algebra makes it possible to talk about spaces of any dimension with identical clarity. Some new concepts live equally well in algebra or geometry: a matrix is associated with a change of unknowns in algebra, and with a change of coordinate axes in geometry.

In the "modern geometry" of Poncelet, Chasles, Plücker, and Cayley, this game of amalgamation reaches new heights. The words they employ are those of geometry—points, lines, conics, planes, quadrics—while the underlying concepts belong to algebra. The idea of a straight line corresponds to that of a first-degree equation, and conics are viewed as equations of the second degree. The object of the game is to take every possible geometric advantage of these algebraic notions, without ever having to perform a single calculation. This exercise of dilettantes, extremely amusing, would greatly influence the reorganization of mathematics. For instance, the transformations by inverse polars invented by Gergonne demonstrate that the notions of point and line are interchangeable. It suddenly becomes clear that the nature of mathematics is more formal than descriptive: *in mathematics, what matters is not the nature of things, but the relationships that exist among them.*

It is also realized that certain geometric properties form a consistent whole with respect to the notions of distance, circle, or right angle. Such properties may thus be seen as a geometry in themselves, given the name of metric geometry. Other properties are invariant under a projection—of one plane onto another, for instance; these are the projective properties. The existence of such families of autonomous properties is elucidated in 1872 by Felix Klein, in his opening lecture at the University of Erlangen. Projective geometry is reduced to expressing, in the form of theorems, those algebraic properties that are invariant under the action of a certain group of transformations of coordinates, the group of projective transformations; similarly for metric geometry, whose theorems express the algebraic relations invariant with respect to a different transformation group—that of coordinate changes from one orthogonal system of axes to another. More generally, a geometry is always associated with a certain group. In such conditions, it is not surprising to see an increasing shift in interest toward the new "structures" that systematize mathematics, and away from the objects themselves—the points, distances, and projections.

And so, as the nineteenth century draws to a close, there is a multitude of new results. They are often deep, and occasionally truly amazing. The borders between disciplines are redefined, and distinctions among some of the most venerable of them all but disappear. The very nature of mathematics changes. From a science that possessed its own traditional objects of study, it is becoming the universal science of relations; in a certain sense the science of the structures that may arise in any science. Such a profusion demanded that mathematics put its house in order.

MATHEMATICS AND INFINITY

As the domain of mathematics kept expanding in search of new conquests, the return to the foundations continued, each of these two movements nourishing the other. Those who are not experts in mathematics may not fully realize today why this double approach was absolutely necessary. This is especially true in regard to the insistence on the purity and universality of the foundations, which is often mistaken for a fixation on logic turned compulsive, while, on the contrary, each stage of the search was the response to some precise—and often concrete—problem.

Let us return, for example, to the question of which functions analysis can, or must, deal with. After the functions that can be easily computed—the first, in order of importance, for eighteenth-century analysts—came as we have seen other special functions. Named after Bessel, Legendre, or others, they can now be found in electricity, earth sciences, and electronics, where their practical usefulness is indisputable. These functions are often defined as the sum of an infinite series, a possibility that Newton had already recognized. One of the first contributions of nineteenth-century mathematics, due in particular to Cauchy, was to determine when such a series can actually be used, that is, the conditions for the series to possess a well-defined sum—or, in technical language, to converge. This is a natural requirement, closely related to the estimation of the error arising from the computation of these functions, a question that no user can afford to ignore.

A function depends on a variable that can normally assume any numerical value, but it is of course impossible to compute the function for the infinitely many values of its variable. Now, if the function has been computed for a certain value of the variable, how much can it possibly change for a new value (of the variable) that is very close to the previous one? It turns out that the functions that can be used in numerical calculations are those for which a small change in the variable produces a small change in the value of the function. These are the continuous functions (which can, of course, be defined in precise mathematical terms).

That was the situation until 1807, when Fourier introduced another way to define and compute useful functions. The series that had been studied until then were so-called power series, that is, roughly speaking, infinite polynomials whose coefficients of higher degree become small rapidly enough for the series to converge. Fourier (or trigonometric) series, on the other hand, are infinite sums of sines and cosines whose oscillations become increasingly stronger. The simplest example of such a series results from decomposing the sound of a musical instrument into its various harmonics—which shows that Fourier's idea was not a worthless curiosity, but a tool physics cannot do without. Fourier series and their generalizations today pervade all technical and scientific fields, and microprocessors are routinely used to compute them at high speed in certain robot components.

It is little known, however, that Cantor's famous reflections on infinity were set off by a problem in Fourier series. These would prove difficult to tame, but they would also be found extremely valuable. In 1830, Bolzano used a series of this kind to construct a continuous function that had no derivative at infinitely many points of an interval. A warning light flashed, because it had always been intuitively assumed that any function that was explicitly defined—even if only by a series—had a derivative. But now Bolzano's result implied that derivatives, those most useful tools in mathematics as well as in physics, could not be taken for granted; that every time the derivative of a function was needed (in the course of a calculation, for instance), one would have to prove first that it existed.

Physicists could afford to ignore such oddities, relying on nature's immunity to that kind of flaw (which is actually not the case). Some mathematicians were also skeptical: "When I come across one of those continuous functions that are not differentiable, I turn away in horror," said Hermite, only half-jokingly. But the truth was that, if physicians must take the Hippocratic oath, mathematicians must honor an unwritten Euclidean oath that obliges them to prove their theorems. They were then forced to prevent that kind of accident by putting up the necessary barriers. (How could they suspect that in so doing, they would be led one day to develop concepts that physics itself would in the end need? But that's a different story.)

This was how Dedekind and Cantor came face to face with infinity. In fact, we find infinity practically everywhere in mathematics. The moment we must take a limit, infinity is there, since the number of terms must necessarily tend to infinity if we are to obtain the sum of a series or define an integral; when dealing with sequences, we have to consider terms of ever higher order, and to define a derivative requires taking values of the variable closer and closer to each other. The notion of a real number—that Eudoxas and Euclid had defined as the result of measuring a concrete length, or any other real quantity—brings in infinity as soon as we decide to express it mathematically (to see this, it suffices to observe that an infinite number of digits are generally needed after the decimal point). Weierstrass and Dedekind preferred to view a real number as the limit of its successive decimal approximations, but this limit is again the culmination of an infinite process.

Geometry cannot avoid infinity either. Euclidean space is infinite, and real numbers (carrying infinity with them) are needed to express coordinates. In each line segment, however small, there are infinitely many points.

One is well advised to keep away from the universe of the infinity-tamers. One of the first, and certainly the greatest, of these was Georg Cantor (1845–1918). It is a temple that should only be entered treading lightly, and where sweeping assertions are banned as meaningless. Infinity was not new: The Greeks had had Anaximander's *apeiron* and the uncountable steps of Achilles in Zeno's paradox. The infinitude of God in all his attributes had also been the subject of study and debate throughout medieval theology and philosophy. But, strange as it may seem, the topic had remained practically virgin, since all the arguments of the past were strewn with paralogisms that no one knew how to resolve. Everything remained to be done, and everything was done, except establish that the final result, "that paradise which Cantor created for us," as Hilbert called it, was indeed the only possible paradise. It was necessary to recognize that the mathematics of today, as abstract, impenetrable, and source of fascination (for some philosophers) or of horror (for others) as it may be, was really the only possible one. It is partly from this fact that the fracture will come.

✣ CHAPTER IV ✣

Classical Philosophy of Knowledge

WHAT DOES IT MEAN to understand? How is understanding possible? These questions, which we shall now address, are among the oldest and most important questions in philosophy. Many answers have been proposed, beginning with Plato's *Theaetetus*, but they can easily be grouped into two main categories. According to the first one, the world is faithfully represented by our mental images and by ordinary language. For the second category of answers, the world is essentially different from our perception of it. It is in fact the opposition between Aristotle and Plato, which has existed since the origins of philosophy.

The birth of science took place in the supposed clarity of the first kind of answer, and the purpose of this chapter is to briefly report its relationship with philosophy during this period. Once again, we shall restrict the discussion to certain facts that will be useful in the sequel.

FRANCIS BACON AND EXPERIENCE

Francis Bacon (1561–1626) deserves to be mentioned first, because he was the philosopher of the experimental method. He anticipated a well-organized, coherent science and, perhaps, was also the creator of a revision of philosophy through science. All this is more important in my opinion than his ideas on method—to which his contribution is too often reduced—exposed in the preface to his *Instauratio Magna*, where he proves an inspired visionary.

He proposes first to build science or, in his own words, to instaur (found anew, reestablish) science. "I entreat men to believe that this is not an opinion to be held, but a work to be done. . . . Now what the sciences stand in need of is a form of induction which shall analyse experience and take it to pieces. . . . I contrive that the office of the senses shall be only to judge of the experiment, and that the experiment itself shall judge of the thing. . . . [I ask you] to be of good hope and not to imagine that this *Instauration*

62

of mine is a thing infinite and beyond the power of man, when it is in fact the true end and termination of infinite error." This "error" that Bacon attributes to Greek and scholastic philosophy was "to fly at once from the senses and particulars up to the most general propositions. [This was] a short way, no doubt, but precipitate; and one which will never lead to nature, though it offers an easy and ready way to disputation."

The instauration of science will be "by no means forgetful of the conditions of mortality and humanity (for it does not suppose that the work can be altogether completed within one generation, but provides for its being taken up by another); and finally it seeks for the sciences not arrogantly in the little cells of human wit, but with reverence in the greater world."

Can one imagine a better description of the future of science, almost at the same moment Galileo was about to put into practice those same principles? If I had to rank the two men, I would favor in one respect the physicist over the philosopher, for Galileo saw that "science is written in mathematical language." Bacon's position is far removed from this conception and his criticism of the possibility of attaining knowledge by logical means is extremely negative, no doubt as a reaction against scholasticism.

The preface to *Instauratio Magna* contains a searing paragraph that I cannot help quoting: "Now my plan is to proceed regularly and gradually from one axiom to another, so that the most general are not reached till the last; but then when you do come to them you find them to be not empty notions, but well defined, and such that nature itself would really recognize as her first principles, and such as lie at the heart and marrow of things."

I shall take the liberty to insert here a personal parenthesis. One of my most precious dreams—partly responsible for this book—is to see one day scientific knowledge so clearly established as to allow a return of philosophy to its pre-Socratic sources, finding in science its own foundations or its most fitting mold. There are times when I believe that such a day has arrived. At any rate, due to that strange tendency to read what we expect rather than what the author wanted to say (and which Paul Valéry has analyzed at length), I once thought that I had found the expression of my idea, forcefully expressed, in Bacon's last citation. Since I had always wanted to know to whom to give the credit (it is so convenient to take shelter under a big umbrella!), I attributed it to him, and so I said (unwisely) in some lectures. I had to change my mind after a second reading of *Instauratio*, because the idea does not explicitly reappear anywhere else. Perhaps Bacon was simply being cautious, and took care not to contradict his own preaching by jumping to premature conclusions. Despite the absence of confirmation, and just in case my first intuition was correct, I like to attribute the idea to him, for want of someone else to honor until much later Husserl came along.

DESCARTES AND REASON

While Bacon's empirical philosophy marked the evolution of science in Great Britain, the rationalistic philosophy of René Descartes (1596–1650) was the respected authority on the continent.

Descartes disagrees with Bacon on one essential point: without denying the pressing need for observation, he nonetheless claims that the prime foundation of scientific enquiry is deductive reasoning. He has had firsthand evidence of this as a geometer, but now it is all of philosophy that he intends to base on human reason, the only secure foundation for the understanding of nature and humanity.

The great avenue of thought that begins with the famous "I think, hence I am" obeys a perfectly clear principle: thought precedes existence, and a reflection using this thought, and on thought itself, offers the method leading to a complete understanding. Reason, more than nature itself, is the starting point.

The reason Descartes should be mentioned here is because he flies the flag of reason higher than anyone else, so high that his "systematic doubt" is clearly a trick. God, whom Descartes claims to meet only later in his book, is already fully present in the *cogito*, ready to occupy the throne reason has reserved for Him. Logic or reason, whatever we call them, are always our most powerful and mysterious servants—or masters?

We shall not embark down the Cartesian road in this book, though, for it is fair to say that Wittgenstein (who many years later traveled along the same path) put an end to it. In his *Philosophical Remarks*, Wittgenstein examines in detail the genesis and the development of language, instrument and precondition of reason. We shall mention only a famous example. A bricklayer is teaching his assistant to speak by naming bricks and tools of the trade. Each time, the apprentice has to point to the object named and say "that." Wittgenstein argues that there is no other way to assign a meaning to the words. Thus, philosophy cannot begin with reason alone, for the latter needs language, which in turn can only become meaningful through direct contact with reality. If the mystery of reason—an obvious prerequisite of science—remains intact, we can search for its origin only in the regularities that Reality exhibits, and not in reason itself. Hence, the *cogito* path is not viable, and, despite its influence on the history of philosophy, we can close it off today without looking back, if not without regret.

Another well-known aspect of Cartesian thought is the procedure explained in *Discours de la méthode* and also, even more explicitly, in *Règles pour la direction de l'esprit* (*Rules for the Right Direction of the Mind*). It consists in decomposing a problem into other, easier ones, until the level of difficulty has decreased so much that the solution imposes itself as obvious. It is perhaps a useful method to guide the mind when reflecting on ordinary, everyday problems. But as a weapon to attack the great questions of our existence, it is practically useless. Such questions refuse to decompose into elementary parts, and the simplicity of their answers, when simplicity there is, resembles more the beginning of a new inquiry than a final conclusion.

We have already mentioned that Descartes viewed matter essentially as extension or, in other words, that for him physics was reduced to geometry. One might try to see in his position an anticipatory vision of Einstein's own conception of matter, but it is rather an example of a resounding failure of the Cartesian method: of the ten propositions concerning collisions that Descartes derived (using his method) only one was correct. In Bacon's words, Descartes wanted to jump to ultimate conclusions in a manner that was both excessive and premature. The question of method in science is more subtle, and we shall come back to it later.

It is undeniable that Descartes left his mark on us all, if only by a certain mechanistic conception of reality, where the physical world and its phenomena are seen as a machine and its various parts—even if the formula is, here again, excessive. We also owe to him, as well as to Galileo, the idea that nature is governed by laws whose form is mathematical—a tenacious, strange idea, that has so absolutely taken possession of scientists that they never even think of questioning its limits. This is what Heidegger called "the Cartesian project": the mathematizing of thought. We shall consider it Descartes' most important legacy and shall often have occasion to return to it.

Finally, we cannot mention Descartes' lineage without recalling at least the great Spinoza. He is above all the philosopher of coherence, and in his company a scientist of our time can be most comfortable. He prefigures perhaps, in more than one way, what philosophy will one day become and even surpass; the day when it will no longer be founded on reason alone—that fragile support—but on a larger knowledge of the *natura naturans* and of the *natura naturata*: Logos and Reality.

LOCKE AND EMPIRICISM

There is no book more pleasant to read and more readily convincing than *Essay Concerning Human Understanding*, published in 1690 by John Locke (1632–1704). Everything in it appears to be obvious and to flow effortlessly, delivered in a crystal-clear language. It could easily be mistaken for the most persuasive application of Descartes' method, if the aim of the book were not, in part, to oppose it.

Locke's thesis is simple: it is the surrounding world that provides us with the means to think and to speak. We all know, says he, what an idea is; be it the product of the imagination or a notion common to all human beings. None of our ideas or principles can be innate, because otherwise small children would already possess them. But, says Locke, small children do not know that it is impossible for something to be and not to be at the same time (that is, they ignore Aristotle's rule of an excluded middle).

This passage, to which other thinkers have replied, is interesting because, for the first time, a philosopher admits having learned something from the observation of children. Genetic epistemology, the systematic study of concept formation in young children initiated in our century by Jean Piaget, confirms Locke's intuition in every respect: All ideas come either from the five senses or from reflective consciousness. It is the concrete objects perceived by our senses that are at the origin of ideas, that is, of the presence, inside us, of their faithful image. Other, more general ideas are the result of reflection, of operations of our mind on those initial ideas. "The soul only begins to have ideas after it has begun to perceive."

According to Locke, the hierarchy of ideas begins with simple ideas, which are not always pure sensory data but the elements that compose them—the idea of ice, for instance, resolves into others, such as those of hard and cold. There is a distinction to be made between the simple ideas that are perceived by a single sense organ, and those combining the data coming from several senses. The latter include space, extension, figure, rest, and motion. At the next stage, the mind extracts similarities among simple ideas through reflection to attain abstraction. In the final stage, that of language, the ideas so formed are described by words, which merely define the common attributes of the things we perceive. And so, everything seems perfectly clear: to understand is to open oneself to the world, whose representation takes form in the mind and generates language and reason. Or is it so clear?

Digression: Cognition Sciences

The subtlcty, or the difficulty, in Locke's approach is apparent in his hierarchy of ideas. The point has been much criticized but its importance is today fully corroborated by science. The example of the idea of ice resolving into those of hard and cold might have been surprising, for it implied that our senses' perception of the external world is a subtle combination of preselected features, rather than a global and faithful image of our surroundings. It is worth insisting on this point, since the question of a human's spontaneous representation of the world is relevant to our main argument: later on we shall oppose a formal representation to a visual one, and so it is important to know what the latter actually is.

We shall refer for that purpose to the present knowledge in brain science, whose progress has been astonishing. Many research efforts converged, a decade or so ago, when several disciplines, up to then independent, pooled their resources to form the cognition sciences: anatomy of the brain, neurophysiology, hormonal biology, biochemistry, experimental psychology. A great number of studies, on both human and animal subjects, contribute to its development, drawing from physiology, and relying on psychological experiments or on the clinical observation of patients who had suffered brain injuries. The observation techniques used are powerful: the positronic camera, for instance, that enables scientists to "see" the circulation of blood through the brain; the nuclear magnetic resonance scanners, able to track the circulation of certain atoms. Computing techniques are of course used for analyzing data, but computer science also enters as a suggestive model of mental processes in many places. An entirely new field of knowledge has opened up, evolving rapidly, and already rich in amazing results.

It had been recognized since Epicurus that perception should play a central role in the foundations of the philosophy of knowledge. Surely enough, its study reveals a strange complexity. Let us consider vision, which is a good example. It was long believed that the almost photographic image formed on the retina is transmitted intact to the brain, where it is analyzed. In this way, the stages of perception and understanding would be neatly separated, but things do not appear to be quite so simple. The retina is a neural tissue of great complexity, which carries out a detailed analysis of the incoming image. Certain specialized neural areas recognize whether the image contains vertical lines, other regions detect horizontal lines; yet others discern colors, the intensity of light, or the presence of motion. The image is thus divided almost instantaneously into various components, which must be recombined by the brain to reconstruct the object seen.

The retina also has a complex network of internal connections, thanks to which it can register and transmit more elaborate information on certain relationships within the image: similarities, but also differences, among the things perceived; the presence of a moving object; and, most of all, that great mystery still unsolved: pattern recognition.

There is continuity from the retina to the brain, since the former is part of the latter. We also know a great deal regarding the different areas of the brain that receive the components of the image. There are four of them, which communicate with one another and react to different characteristics of the image: motion, color, and form. This last component represents information so rich and complex that its processing requires two areas of the brain.

The synthesis of a full perception involves even larger regions of the brain. Those that are active at a given instant can be directly observed with the aid of the positronic camera. The camera shows which parts of the brain are being activated by displaying the flow of blood through it. In the study of vision, it is possible to correlate these data with the object seen by tracking the movements of the eye with a video camera. Experiments of this kind have been performed on animals by implanting directly in their brains electrodes that receive the neural signals. This technique has revealed how different areas of the brain react, depending on the characteristics of the sight before the subject's eyes: color, motion effects, changes in form, the presence of recognizable shapes, and so on. The coordination of the process takes place in the cortex prefrontal lobes, where the short-term memory is located. The latter can detect, for instance, the appearance of a new element in an otherwise unchanged scene. It is also from this region that commands to move various parts of the body are sent out: eye, hand, or mouth movements. The prefrontal lobes are therefore able to control attention and the ensuing reaction.

Thus, perception appears as an extremely complex process, where the outside world is first decomposed into a multitude of attributes, well before its meaning can be grasped. This analytic feature of perception, which begins by breaking down the image before proceeding to its synthesis, has gradually imposed itself during the last few decades, and philosophy can no longer ignore it.

This long digression was at least useful for stressing a point which, already noticed by Locke, has presently become somewhat more cogent: Our mental representations, even if they originate from the world around us, are reconstructions. They are far from being simple or obvious. Their validity is most questionable when science takes us to unfamiliar surroundings, among electrons or the universe as a whole, for instance. One may therefore wonder whether clarity is not, after all, merely familiarity, and, outside these familiar surroundings, only a semblance.

HUME'S PRAGMATISM

David Hume (1711–1776) is descended, philosophically speaking, directly from Locke. He reduces the totality of knowledge to a product of experience, and is even more radical than his predecessor in denying any significance to questions beyond the range of experience. He particularly impressed his contemporaries and marked the history of philosophy by his absolute rejection of any metaphysics, to the point of negating the existence of universal moral principles, and even moral concepts, such as liberty. This aspect of his philosophical writings was naturally the most controversial, but we shall not be concerned with it. We shall only discuss Hume's views on human understanding and on the nature of science.

His *Inquiry Concerning Human Understanding* appeared in 1748, a sequel to *Treatise of Human Nature* (1737), written during his youth. *Inquiry* contains the most explicit—and practically definitive—formulation of his pragmatic philosophy, still widely followed today in scientific circles. Let us hear Hume himself sum up his main thesis: "But though our thought seems to possess [an] unbounded liberty, [one finds], upon a nearer examination, that it is really confined within very narrow limits, and that all this creative power of the mind amounts to no more than the faculty of compounding, transposing, augmenting or diminishing the materials afforded us by the senses and experience." The only function of our mind is therefore to exploit *facts*. Reflection does not add anything of substance to the information provided by the facts. Most notions of speculative philosophy thus become meaningless, as a simple test shows. Indeed, it is enough to ask the question: From which sensory impression is that supposed idea derived? If we cannot answer it, we were talking of a notion devoid of any meaning.

Next comes something more relevant to our discussion: the laws that science discovers through experience merely reveal the existence of a customary conjunction among facts (custom being thus the great guide of human life and understanding). *Scientific laws are simply a summary of observed facts.* Hence, facts are the source of our representation of the world, of our language, and this is possible because facts have enough regularity to allow reason and language to be useful. This regularity is best described by the laws articulated by science, but these laws do not add anything to a bare summary of facts.

At the end of this scouring exercise where mind is reduced to a lame copy of the world, it is surprising to see Hume betray without warning his own rules and suddenly become a radical metaphysician. This is at least my interpretation, if metaphysics consists in decreeing what the world must be, rather than accepting what it is. To be sure, he admits that there are intimate connections among facts, which the laws reveal, but he claims with absolute certainty that it should be impossible, inconceivable, to learn anything else in that respect. He condemns our logic as definitely helpless, its inferences being only some habit grown of mimicking facts that repeat themselves endlessly.

This aspect of Hume's thought is extremely important and was to become one of the hottest issues in the philosophy of knowledge. A great number of questions are declared off-limits and forever unanswerable—if not meaningless—by Hume, even in the sole domain of science: Why are facts related by the consistent patterns that experience reveals? How come that the "laws" of nature, such as those of Newtonian mechanics, enable us to predict the result of experiments that have never taken place, and where does this predictive power of science come from? Hume rejects all these questions as being beyond what we can ever hope to know. The gradual discovery of the fact that we sometimes can go beyond this limitation will constitute, as we shall try to show, the true domain of today's philosophy of knowledge.

KANT

In his *Biologie des passions*,[1] the well-known neurobiologist Jean Didier Vincent marveled at the exceptional quality of Spinoza's observations on the workings of the human mind, which are often in close agreement with the most recent and significant results in neuroscience. Vincent's remark could also apply to some of the most important philosophers of the Enlightment and earlier, who were undoubtedly subtle and acute psychologists. These include Descartes (in his *Meditations*) and Malebranche—as well as Locke and Hume, it goes without saying. As far as Kant is concerned, it seems that a more radical statement can be made, namely, that his main contribution to the philosophy of knowledge should rather be considered as pure psychology. Immanuel Kant (1724–1804) stands among the most important philosophers who ever lived, if

[1] Paris: Odile Jacob, 1986.

only for his depth and his exacting rigor. Though strongly influenced by Locke and Hume, he could not accept the latter's renouncing further investigation into the origins of the laws of nature. This question was for Kant a *crux metaphysicorum*, as he called it. He expressed it beautifully in the famous first sentences of the *Critique of Pure Reason*: "Our reason has this peculiar fate that, with reference to one class of knowledge, it is always troubled with questions which cannot be ignored, because they spring from the very nature of reason, and which cannot be answered, because they transcend the powers of human reason."[2]

But what is reason? Kant's definition—the faculty for producing unity among the rules of understanding according to principles—is of little help. It comes only at the end of a long construction requiring for its validity the full acceptance of the *Critique*'s basic assumptions. The significance of reason in modern cognition sciences would rather take into account two simple but essential ideas that were still missing in the eighteenth century: that thinking takes place in the brain, and that reason, and logic in particular, was acquired and developed by humankind through a long historical process. It has a strong social component, involving communication and culture.

Kant does not refer to the brain, and he was certainly right in doing so, given the little that was known about it in his time. An interpretation of Kant with reference to the brain will be convenient, however, even if reductive, in a necessarily short discussion. It at least makes clear the distiction introduced by Kant between two kinds of knowledge, *a posteriori* and *a priori*. A posteriori knowledge is easy to understand since it means everything we draw from empirical intuition (*Anschauung*, the German word, being certainly the most accurate). Kant is essentially concerned with something that exists before any intuition can take place. A convenient way of interpreting this a priori datum would be to consider it simply as the framework and the functions of the brain.

A basic aspect of Kant's approach is its emphasis on the nature of *phenomena*. A phenomenon is the "undefined object of an empirical intuition," something we could rather try to define as a state of the brain when it perceives an external thing or an internal bod-

---

[2] The translation is by Max Müller, as quoted in Walter Kaufmann's *Philosophic Classics*, vol. 2 (Englewood Cliffs, N.J.: Prentice-Hall, 1968).

ily process. Our present understanding of perception agrees with the need to distinguish real things from their representations through perception, or reality from "phenomena." We cannot, however, overlook the fact that our brain obeys the laws of physics, chemistry, and biology. Either we accept Kant's idea that nothing can be known except phenomena and follow him (and later Husserl) down the path of phenomenology or (except for joining Hume in his renunciation) we recognize in the crux of metaphysics an urge to investigate more deeply the laws of nature and their stupendous unity.

Kant sees the organization of phenomena as dominated by two basic "a priori synthetic judgements," which are space and time. They are a priori because they belong to the domain of reason and not intrinsically to the objects we can observe (or at least this is what Kant says). They are "synthetic" in so far as they are concerned with the ordering, the synthesis, of various phenomena. For instance, different objects are placed side by side in space *by the mind* and not mixed together; different events in time are similarly ordered in succession. Our intuition, our awareness, of the world outside and inside us is necessarily cast into these molds of space and time, which are supposed to have no external, genuine reality.

The status of space and time is an essential element of Kant's philosophy, and is also a paradigm for all modes of thought. In the *Prolegomena to any Future Metaphysics*, which is a sort of "popular" version of the *Critique*, Kant makes this basic point very clear: "Understanding does not derive its laws (a priori) from, but prescribes them to, nature." This radical change of reference for the generation of knowledge is often compared by the authors of philosophy textbooks to the Copernican revolution. Kant offers it as a way of breaking through Hume's crux of metaphysics: Reason is enough to explain the existence of laws and the regularities of the world!

In order to appreciate the value of Kant's ideas in the present day, it may be interesting to compare his approach with some of the formal sciences that have emerged in the meantime. The notions of space and time immediately suggest a comparison with the theory of relativity. We may notice first that neither Kant nor his followers fully exploited the possibilities of their assumptions. They took for granted that different observers, different intuitions, would necessarily have to agree in their synthetic a priori judgements. There seems to be a slip in the reasoning here, because such a comparison belongs to the empirical domain, whereas a transcendental, a priori analysis should have considered the various alternatives, that is, a possible agreement or disagreement between the representations of different observers. In that sense, Einstein may be said to have carried Kant's arguments farther and in a more careful way than they had previously been. In other words, relativity does not directly invalidate the principles of Kant's approach, as Ernst Cassirer observed.

To find out where the conflict really is we must look at a later part in the *Critique*, when the famous antinomies are discussed. These are four pairs of opposite theses that Kant asserts to be forever undecidable by reason. He offers proofs of this claim using "transcendental" methods (that is, the principles of pure reason) although most of these "proofs" are not more convincing than many in Aristotle's works. The fourth antinomy has to do with the existence of God and the third one is concerned with causality and free will; hence, neither of them is relevant to our discussion. The second antinomy is more directly linked with science, since it claims as undecidable whether a compound substance is made of simple parts or not. In other words, the existence of atoms is accepted as a sensible assumption, but it is shown that it will never be possible to prove it. At this point, Cassirer notwithstanding, one feels that something wrong has crept into the argument, though we shall not attempt to discuss what and where.

In the first antinomy, two theses are opposed: the first one states that "The world has a beginning in time and is limited also with regard to space," while the other maintains that "The world has no beginning and no limits in space, but is infinite in respect both to time and space." This may be compared with the opinion of most modern cosmologists who claim to possess solid arguments indicating that the universe had a beginning, and consider that the finiteness (or lack of it) of the universe can in principle be decided by accurately measuring the average density of matter in space. One is left wondering how these scientists can maintain that they have solved what was considered as definitely undecidable by Kant. The essential point is that the cosmologists rely on a mathematical concept, space-time—together with some physical laws it obeys—that clearly falls outside the range of intuition. Since it does not belong to the system of Kant's phenomena, it is not subject to his judgments.[3]

It turns out—though of course, both Hume and Kant could not help ignoring it—that some sciences can deal with facts that are accessible neither to intuition (*Anschauung*: the vision of what is offered immediately) nor to perception. The existence of these "formal" sciences, which we shall consider in more detail later, is therefore an essential question (and a valuable hint) for philosophy. Kant has nothing to tell us that would help us understand formal sciences. He is, however, extremely helpful and reliable when we try to pinpoint exactly what makes these sciences formal, when and where they depart from classical thinking, and how they extend it. Kant provides us with the acme of this way of thinking, this "pure reason," and he is by far the deepest analyst who ever existed regarding everything intuitive. Though we cannot accept his statement that reason prescribes its laws to nature, we may rely on him for perfectly delineating *classical* thinking: thought proceeding uniquely through intuition, vision, and unformalized clarity. Kant is an irreplaceable reference for the definition and recognition of classicality, in view of his insightful psychology of rational consciousness. This is why we proposed an unusual evaluation of his work that at first might have appeared improper and controversial, but whose legitimacy should by now be clear.

Kant's analysis extends to other important aspects of classical thinking. His categories of understanding, in particular, provide a careful and systematic review of the modes of reasoning (when reason intends to use intuition, and intuition only, as a source of information). Among these twelve categories, those of Reality and Causality deserve special attention. One may skip many long discussions of these difficult categories in the modern literature and go directly to Kant's book to find out why reality and causality intrinsically belong to classical thinking. This will be important for fully appreciating the absence of these categories in quantum

---

[3] To avoid an apparent contradiction between what is said here and what we said before (about the absence of contradiction between relativity and Kant's considerations) it should be stressed that in the special theory of relativity the concept of space-time is convenient but not necessary. The tentative cosmological answers to the first antinomy rely on the relativistic theory of gravitation (general relativity), in which space-time is a fundamental concept.

physics. We may also notice that Kant's category of inherence (if I understand it correctly) agrees with Leibniz' discussion of indistinguishable objects: two objects (two substances) should always be distinguishable by some feature in classical thinking. Quantum physics gave up this notion of inherence also in one of its most basic principles (Pauli's principle).

Several conclusions may be drawn from this too short analysis. The main one was already mentioned when we discussed Bacon. Considering how far from Kant's schemes science has brought us, and recognizing that nothing as systematic as the *Critique of Pure Reason* has been produced in the meantime, the need for a new foundation of philosophy is clearly imperative. Husserl said it earlier, but unfortunately he became muddled with Kant's legacy of phenomenology.

Understanding is a human quality—though some people may envision other kinds of understanding for machines or aliens. Kant's psychological lessons on understanding should never be left out of any search for new foundations. Since he did not take the brain explicitly into account, he could not appreciate that there is always an experience prior to any a priori—transcendental—act of the mind. We know, on the contrary, that there is an ontological construction of each human brain since infancy as well as an on-going phylogenetic evolution of *Homo sapiens*. This is where Kant's error lies, and where his heroic attempt at solving Hume's crucial problem failed.

Does this mean that a true foundation of philosophy, which should certainly be at least as ambitious as Kant's, will have to wait for a total understanding of the brain? Probably not, if we do not aim at an impossible completeness and take over the task where Kant left off. We are in a much better position, because what we must investigate is no longer Hume's problem, but how and why this metaphysical crux has been overlooked by science. We shall see that it is probably because the roots of logic, if not of reason, are not to be sought within the structure of our mind but outside it, in physical reality. This of course implies a complete reversal of Kant's approach.

Our brief review of philosophy must now come to an end. Though outrageously sketchy, it will serve as a background for our endeavor, if only because the ideas of the past still impregnate our minds through culture and teaching. If we are to overcome them, better to make them explicit. In any case, this review has at least shown us how urgent, but also how difficult, a constructive philosophy of knowledge can be.

The story does not, of course, stop at Kant. There were other great psychologists, such as Nietzsche and Freud, but they have little to teach us that is relevant to our enterprise. Others have attempted to construct a theory of knowledge, and we may mention those of Bertrand Russell (1872–1970), Alfred Whitehead (1861–1947), Ludwig Wittgenstein (1889–1951), and Edmund Husserl (1859–1938). All of them, in some sense, were born too late and too early: Too early to seize the full implications of recent scientific discoveries—in particular the laws of the quantum world—and too late to prevent the abrupt collision of their views with the new insights. It might well be, under such constraining historical circumstances, that the greatest philosopher of our age was Niels Bohr, but to say that in the present chapter would be anticipating too much.

PART TWO

THE FRACTURE

✢

THERE IS NO DOUBT that we are going through a period of fracture, whose first perceptible manifestations are four centuries old, dating back to the dawn of modern science. But if fracture there is, what is it exactly?

There are at least two aspects of this fracture whose significance is widely recognized. The first one concerns humankind's place in the universe and our perception of it, while the second has to do with the pervading consequences of modern technology. The former aspect influences the mind, the latter affects life as a whole.

Who can ignore today that we are part of the evolution of all species, of the universal flow of life; that the formation of the sun and the earth extends our lineage even further back in time, right down to matter, to the oxygen breathed by the first living creatures, to the atoms that compose us and which once were part of long-dead stars—that the universe had a beginning? I would assume that we all know these things, and that they are the intellectual background of our century. As for the proliferation of technology, our earth being crisscrossed by airplanes, waves, and information, the changes in our daily life, the impact of medicine, and the cries of a world in pain and suffering, we know how important they are, but, again, these things are well known and I have nothing to add to the books and essays that discuss them.

I would like to talk of a fracture that is more discreet and little noticed, but also important. It concerns a profound transformation of science, one that measured on a historical scale has just occurred, and which greatly affects the nature of thought, the act of understanding. It takes part in an eminently positive movement, that powerful trend toward coherence and order we mentioned in the Preface, where the laws originating from each particular science come together to form a seamless bundle of imposing unity. The fracture is nevertheless there, in the fact that these laws are, when seen through the eyes of average intelligence or classical philosophy, absolutely incomprehensible. In a nutshell, the more we know, the less we seem to understand.

We often hear the legitimate complaints of those who cannot understand the principles of contemporary physics or mathematics, which no amount of "popularization" succeeds in communicating. There is in this situation more than meets the eye, more than the consequence of an excessive specialization or an immoderate taste for abstraction: the existence of an intrinsic darkness.

It is even worse, as we shall see, and the traditional foundations of philosophy now give up under the pressure of science. It is impossible to describe this breakdown in a few words, for it does not seem to have been recognized in all its ramifications. Let us just say that we are losing the spontaneous representation of the world that used to be at the origin of every thought; common sense is defeated together with the philosophical principles it generated. A strange predominance of abstractness, of formalness, exists at the very heart of reality. There can be only one remedy: to invent a new way of understanding.

The cracklings announcing the fracture were clearly heard, but their deep rumblings went unnoticed, and it is under this incomplete form that they already traverse philosophy. First, there was a retraction of logic on itself, when it becomes formal and introspective. The books are full of it, from Russell to Wittgenstein and from Carnap to Quine or Popper. Formal logic helped another, wider renewal in mathematics, cutting the last ropes that still tied it to reality. Mathematics became autonomous, a pure game of relations, Logos renewed, where Forms were no longer forms of something concrete but ready to accommodate anything. Many authors discussed it, those mentioned above as well as others, such as Jean Cavaillès and a number of our contemporaries.

The major upheavals took place in physics: First, the theory of relativity and its questioning of the categories of understanding theorized by Kant; then, and especially, this almost universal science called quantum mechanics, which is in fact the general expression of the laws of nature in a world made of omnipresent and almost imperceptible particles. It is this science that warned us of the limits of common sense and the fallibility of some fundamental philosophical principles: intelligibility, locality, and causality, for instance. Words fail us; they only encompass the most deceiving appearance of things, and bump into each other in multiple contradictions. Only mathematics has the fiber to capture the concepts of physics; not merely to render them precise, as in the science of the old days, but to articulate them—and nothing can replace it.

We have just outlined the essentials of this second part: a diagnosis of cataracts, the clouded vision of science, be it in formal logic (which we shall barely touch upon), contemporary mathematics, or quantum mechanics. We shall also recall in broad outline the acute philosophical perplexities that arose in connection with those sciences, be it in mathematics or in quantum epistemology.

In so doing we are preparing the ground for a last stage, to be addressed in the final part of the book but requiring this preliminary analysis. Any attempt to renew the philosophy of knowledge at a level suitable to the complexity of the current problems cannot be supported by a collection of disjoint reflections—a bit of logic here, of mathematics or physical sciences there—appearing in so many separate and specialized books, as is now the case. The keystone should rest on all three of these pillars at the same time, however unhappy the specialists might feel. This is the reason we build them now.

A last remark, concerning the terminology we shall adopt to better emphasize the philosophical characteristics common to these three sciences, and to identify the two major stages of their evolution. We have referred to physics before the fracture as "classical" science, and to mathematics afterward as "formal" science. Thus, Aristotle's logic is classical, as is Newton and Leibniz's differential calculus; while quantum physics, despite the number of its very concrete applications, will be called formal. This is simply a particularly convenient classification to render our argument more transparent.

Formal Mathematics

THE AGE OF FORMALISM

Fᴿᴏᴍ ɴᴏᴡ ᴏɴ, we shall cease to follow the unfolding of the history of mathematics, which, by the way, is accelerating,[1] and shall rather focus on three ideas we had already come across. The first one concerns the very nature of mathematics, whose purpose is not specifically the study of numbers, geometric figures, or any other particular domain. These are only applications of mathematics and not the essence of it. This essence is the study of the relations that exist among concepts, irrespective of their specific nature. In short, it is the abstract study of form.

The end of the nineteenth and the beginning of the twentieth century are characterized by a massive effort to conquer the formal domain, to stake it out, establish its rules, and draw up its map. Such an enterprise requires an extreme caution in order to thwart habits or laziness of thought, beginning with a merciless hunt for deceiving appearances. These are the reasons for the army of symbols that takes over the field of thought. Any temptation to yield to the lure of intuition is strictly banished. Hilbert went as far as proposing, half-jokingly, that suggestive words such as "point," "line," and "plane" ought to be abandoned, and replaced with "table," "chair," and "glass."

Such an intense preoccupation with formalism and this mistrust of intuitive representations would necessarily have an impact on logic. The latter could no longer afford to deal only with the clear but superficial images suggested by the mind. It had to become a logic of formalness, a formal logic.

---

[1] This chapter is based on the collective work *Abrégé d'histoire des mathématiques 1700–1900*, edited by Jean Dieudonné, 2 vols. (Paris: Hermann, 1978), and on the collection of original articles edited by Jean Van Heijenoort, *From Frege to Gödel, a Source Book in Mathematical Logic, 1879–1931* (Cambridge, Mass.: Harvard University Press, 1967).

The second idea concerns the importance of infinity in mathematics. Infinity is everywhere, except in a few precise domains. It is also one of the most difficult concepts for logic to handle, and requires a corresponding evolution of the latter.

The third important idea is the search for consistency. When analysis was reformulated to satisfy Euclid's logical standards, it became necessary to reexamine first Euclid's own axiomatic model. The role of the hypotheses was clarified. An axiom was no longer a self-evident truth, nor a postulate an assumed one; both were rather viewed only as possible truths, worth exploring for the profusion of their logical consequences.

But in cutting itself loose from physical space and other such references to the concrete world, mathematics also lost the certitude—actual or presumed—that reality guarantees. New, unexpected obstacles, hidden errors, and internal contradictions then became a threatening possibility. This state of affairs prompted a new reflection on the notion of mathematical truth, which was now to be understood as a freedom from contradictions, a complete logical consistency.

This whole evolution toward a formal approach proved amazingly prolific, and not, as one might have feared, an empty shell. It opened countless new domains inside mathematics itself, and, strangely as it may appear, far from consecrating a divorce from reality it established a new reunion with it. Indeed, shortly after the transformation of mathematics, physics would in turn be propelled toward the new landscapes of relativity, of the relativistic theory of gravitation, and of quantum physics, and some of the boldest constructions of formal mathematics would be found to be indispensable in the formulation of the laws of nature. Nothing seems to explain this astonishing encounter, but it is undoubtedly a *philosophical discovery* our age has yet to fully recognize.

This formal, present-day mathematics will be our next topic. We do not wish to get lost in niceties or technicalities, nor, unfortunately, give in to the temptation of some of its beautiful sights. In a certain sense, we shall resemble Saint Bernard meditating on his way to the Alps without even glancing at the majestic peaks, absorbed in what he considered to be essential.

FORMAL LOGIC

The person who did the most to revive logic in our time was unquestionably Gottlob Frege (1848–1925). The need for a revival was felt toward the end of the nineteenth century, when infinity had to be tackled head-on for mathematics to move forward. There was, of course, some resistance against the necessary increase in the level of abstraction, by the same kind of people who today marvel at the achievements of computers, which are really machines to do formal logic.

It is difficult to do justice to either Frege's or Cantor's far-reaching contributions in a few sentences. We must nevertheless say something about logic, if we wish to discuss present-day mathematics, and we shall encounter it again when trying to understand quantum physics. But we need not embark on the analysis of Frege's difficult works, and a few words on those of his predecessor George Boole (1815–1864) will suffice.

Boole's principal achievement was to realize one of Leibniz' dreams: to devise a practical symbolism, together with a complete set of reliable rules, for carrying out the logical operations in a simple and automatic fashion. One of his prime ideas was to replace the definition of the properties of an object by a reference to all the objects possessing those properties. For instance, instead of defining "red" using words which would inevitably prove inadequate, he supposes that one can always decide whether a given thing is red or not; all red things then form the set of red objects. Boole therefore favors extension, or the set approach, over comprehension, which requires precise definitions. Similarly, it is assumed that the set H of all humans and that of all mortals (M) exist. The statement of a simple proposition such as "All humans are mortal" can then be expressed by saying that the set H is contained in the set M.

Boole also proposes a simple and elegant symbol for the logical conjunction "and," which he denotes as a product. Given, for instance, the set B of black-haired humans and the set F of all females, the set of black-haired women is then denoted by $F \cdot B$; it is the set of elements common to both F and B. In a similar fashion he defines the logical alternative "or," denoted by the addition symbol. Females *or* black-haired humans make up a set called the union of the two sets F and B and denoted $F + B$ by Boole (to be fair, this inclusive "or," different from the exclusive "or" that corresponds to "this, or else that, but not both," was not considered by Boole himself but introduced by Jevons).

Another of Boole's ideas was to fulfil a need that logicians sensed but had failed to pinpoint: the need for universal sets. One of these is the empty set (the set with no elements), which Boole denotes by 0. The other one is the "universe" to which any given logical argument implicitly refers—the previously mentioned universe of discourse or *Denkbereich*. We can illustrate this notion with an example. Suppose that we are talking about marriage. This necessitates the notion of married people, that is, the set of men and women that are married. After that, we could consider the set of married couples, but which marriages are we talking about? Monogamous, polygamous, or polyandrous (a woman having several husbands, as existed not so long ago in Tibet)? These are three possible universes of discourse, and the structure and properties of the underlying set would be very different in each case. Logic requires that we specify, before everything else, the relevant universe of discourse or universal set of reference. Boole represents this "universe" by the symbol 1.

Negation can then be described in terms of sets as well. Boole views a proposition as asserting that certain elements belong to certain sets. For example, if the universal set 1 is the set of all humans, one element of this set is Socrates, and the set B of all blond humans is contained in 1. Now, the proposition "Socrates is blond" expresses precisely the fact that the element Socrates belongs to the set B. The negation of this proposition ("Socrates is not blond") amounts to introducing the set B' of the elements (of 1) that do not belong to B, and it says that Socrates belongs to B'. The set B' is denoted $1 - B$ by Boole.

The above logical symbolism works well as long as it is handled with care. In other words, the manipulation of the symbols (1, 0, $\cdot$, +, −) is governed by precise rules that resemble, but are not identical to, those of algebra. Going back to the sets B (blonds) and F (females), we clearly have $B \cdot F = F \cdot B$ (for every blond female is a human of female sex who is blond). But we also have $F \cdot F = F$ (every female woman is a female). The precise rules of this algebra of logic were clearly stated by Boole and later completed by A. De Morgan and C. S. Pierce. They can be found today in every textbook on logic, including those intended for future electrical engineers and computer scientists.

Next comes logic's master key, the one that allows a statement of the consequences of given premises: logical implication (already known to the Stoics). In Boole's system, a proposition *a* implies another proposition *b* if the set *A* corresponding to property *a* is contained in the set *B* corresponding to *b*. This can be conveniently written as the "equation" $A \cdot B = A$.

More generally, Boole's works have the merit of establishing the close connection between logic and set theory. One may not totally agree with his definition of a property in terms of a set—and soon we shall see an alternative one, using symbols, as Frege did. But logic nevertheless always requires some set of reference, a collection of all conceivable propositions making up the universe of discourse.

For a while, a predictable question came up: Can all of mathematics be simply reduced to logic? The two thick volumes of Russell and Whitehead's *Principia Mathematica* are based on a positive answer to that doctrine. A few years later, Bourbaki chose rather to use set theory as the foundation of mathematics, and assign logic a secondary role. Which one is the true beginning? A clear answer has yet to be given—and is probably impossible to find.

Symbols and Sets

The little we have said about logic is assuredly insufficient, but our real objective lies elsewhere. It is to give an idea of the nature of contemporary mathematics, with its formalism as intrinsic as it is arrogant. In order to describe this science of symbols and relations we shall begin with the symbols.

All is based on the principle of the excluded middle in its purest form: in the beginning there are two *distinct* symbols, denoted 0 and 1. We could have employed any other symbols, a circle and a cross, say, or—had this book been printed in color—a blue dot and a red dot. At any rate, what is important is to realize that here 0 and 1 are not numbers but symbols. But symbols representing what? Representing only themselves. Give me 0 and 1, a present-day Descartes would say, and I will give you thought. Without going that far, let us say for the moment that we are only interested in saying what a symbol is, and that only two of them suffice to generate them all.

Let us assume, for convenience, that we possess a computer. In particular, our computer has a memory in which we can store elementary symbols such as 0 and 1 (the binary digits, known familiarly as "bits") or strings of them. Each of the symbols 0 and 1 is physically implemented in the machine by an electric potential between the terminals of a transistor, a potential that can take on two different values. The computer's memory is divided up into various units, each of which will be used for a particular purpose. The computer being a finite machine, it has a finite—if possibly very large—memory capacity, just like our own brain.

Starting with the elementary symbols 0 and 1, other (composite) symbols may be constructed, simply by repeatedly writing the former in sequence. The result is a string of 0s and 1s, such as 0, 1, 10, 11, 100, 101, 110, 111, 1000, and so forth. The length of each sequence is finite, and so is the totality of them that can be stored in our (finite) computer. With these symbols we can express a variety of ideas: write numbers, represent points, circles, logical or arithmetical operations, and so on.

Even if the strings of 0s and 1s are sufficient for us to build from them all of mathematics, our mind, unlike the computer, is not at ease with their monotonous simplicity. For this reason it will be convenient to discuss mathematics at two different levels, one describing it as it really is and the other as we humans like to hear about it. In the first version (for the computer), everything is written and expressed in terms of the two elementary symbols. Such a language is well suited to a perfectly abstract thought, for its very starkness is a safeguard against unwarranted interpretations of the symbols by our imagination. But, on the other hand, such a purified language soon becomes incomprehensible to our mind. And so we will occasionally talk "between ourselves," that is, between humans, or between the author and the reader, in ordinary language.

Our first task will be to teach elementary set theory to the computer. But we shall not proceed along the lines of Boole's algebra of logic, because despite its apparent simplicity it is beyond the computer's reach, or, put differently, it is still too intuitive to be formalized. The set of dark-haired women represents something for us—an image, an idea—but it means absolutely nothing to the machine, which does not understand words, in particular, the word "set." We must begin at a deeper level, and teach the machine the grammar, so to speak, of the theory, without any reference to meaning, to intuitive representation, to semantics; in short, we must teach our computer the *formal language** of the theory.

Each unit of memory will play a different role. There is the *Elements* unit, the *Set Names* unit, a *Sets* unit, and also *Signs* and *Propositions* units. There may be other units as well, but we have enough for our purpose, which is to illustrate the formal method. The *Elements* unit contains a fixed collection of symbols—that is, of strings of 0s and 1s. We shall designate them "between ourselves," by lower-case letters: *a*, *b*, *c*, etc. In the *Set Names* memory unit we include a symbol that is supposed to name the set stored in the *Elements* unit. This set, denoted by *E* between ourselves (as a rule, we shall employ upper-case letters to designate sets) may be viewed as the *universal set* (or, in Boole's notation, 1). We are going to use it to teach set theory to the computer in a practical way, that is, by telling it the rules for handling the different symbols.

The next stage consists in constructing all the subsets of *E*—quite simple, really. First, the one-element subsets; this amounts to copying the contents of the *Elements* memory into the *Set Names* memory. Now, since they have changed memory unit, the former "elements" no longer represent the same concept—they have now become what mathematicians call "singletons," and are written, for instance, {*a*}, to designate the set whose single element is *a*. In a similar fashion we obtain names {*a*, *b*} for all subsets of two (distinct) elements, then the three-element subsets, and so on, until all the subsets of *E* have been named, including *E* itself. For good measure, we throw in another symbol among the set names: the empty set or set with no elements, usually denoted by ∅ (0, in Boole's notation).

In the *Signs* memory unit we store only four symbols, denoted (between ourselves) $\in$, $\notin$, $\subset$, $\not\subset$. They represent certain relations between elements or sets that we express in words by belongs to, does not belong to, is contained in, is not contained in, respectively. For example, we write $a \in A$ to indicate that the element *a* belongs to the set *A*. The computer can "understand" the sense of this proposition by checking whether the element *a* occurs among those of the set named *A*. If such is the case, it will then be instructed to store in the *Propositions* unit the sequence of symbols $a \in A$; otherwise, it will write $a \notin A$. Similarly, we say that the set B is contained in the set A, written $B \subset A$, if every element of B also belongs to A. Again, the computer can verify this assertion and store $B \subset A$ or $B \not\subset A$ (whichever is the case) in *Propositions*.

But, the reader may be wondering, what are the assumptions regarding the operations that the machine can execute? The answer is that the computer obeys the basic laws of formal logic established by Frege and Peano (we shall not, however, discuss them here, and simply acknowledge this fact). Apart from that, our computer is not essential to the construction, it is just a convenient device to emphasize the purely symbolic character of logic, and the total absence of any visual representation.

The above preliminaries may appear slow and tedious, but we hope that they nevertheless show how the concepts introduced so far can be expressed by means of symbols only, without the help of any underlying intuitive representation. We can see in this example how a mathematical theory may be systematically built as a logical organization of symbols, the meaning of which need not be specified in advance. The "elements" can be names of students of some university, and a certain subset may represent the football team; but the elements may just as easily be the fruits of an apple tree, and the subset the contents of a basket. In the universe of symbols, the only meaningful facts are the relations among them.

PROPOSITIONS

We shall now move on to a higher level of abstraction. An important feature of mathematics is its ability to deal on an equal footing with things that are real and things that are only possible. A line segment may equally well be the one connecting the tops of two real mountains, such as the Acropolis and Lycabeth in Athens, or simply a possibility, as when we say, let AB be a segment. Likewise, when speaking of a number, it could be the number 8 or a possible number n, of which we know nothing. It is precisely this versatility of thought that we shall communicate to the computer, our guinea pig in formal thinking.

Instead of speaking of some objects as being real (or explicit) and others only possible, we shall call the former concrete and the latter abstract. To accommodate these two kinds of objects we shall need some memory space. The former *Elements* unit will now be divided into two sections, *Concrete Elements* and *Abstract Elements*, and similarly for *Set Names*.

To illustrate the idea, assume that the elements of E are the symbols 0, 1, and 10 (called 0, 1, and 2, in ordinary language). These would be stored in the *Concrete Elements* memory unit. In order to be able to speak of a generic element, or "element a"—the "number x," as one would say in algebra—we choose a symbol to represent the letter a and store it in *Abstract Elements*. If in the course of a logical argument we find that $a = 2$ (or if $a = 2$ is one of our hypotheses), it would then be easy to connect the memory cell in *Concrete Elements* containing the symbol 10 (the number 2) to the memory cell in *Abstract Elements* where the name a is stored.

The signs, such as $\in$ or $\subset$, can also be employed to express relations between abstract elements or sets. Here's an example: If P denotes the set of even numbers, then the proposition $6 \in P$ can be readily verified by our computer (assuming it can perform elementary arithmetic), for one can instruct it to check whether 6 is (exactly) divisible by 2. But the proposition $a \in P$ cannot be so verified, unless we provide the machine with additional information. We are going to introduce now two important notions: those of an abstract proposition and of *metalanguage*.*

A formal proposition is actually a sequence of symbols constructed according to certain syntactical rules. For instance, by the proposition $a \in A$ we understand that the element a belongs to the set A. This entails that the first symbol (a) must belong to the list of element names, abstract or concrete, that define A; and this symbol A must be a (concrete or abstract) set name. As for the sign $\in$, if the sequence of symbols is to make sense, it can appear only between an element name and a set name. There are many other analogous symbols and signs in mathematics which must be combined in a suitable manner for the resulting sequence (or formal proposition) to have a meaning. In elementary set theory, some of these other signs are $\cap$ for the intersection of two sets, and $\cup$ for their union (originally denoted $\cdot$ and $+$ by Boole). The rules governing the writing of "sentences" constitute a kind of grammar that our computer must learn.

One of these grammatically correct sentences is $((a \in A) \cdot (a \in B) \Rightarrow (a \in A \cap B))$, which means (in our language) that if an element a belongs to a set A and also to a set B, then it belongs to the intersection of these two sets. We humans may have derived such relations between sets from observation, but this fact is irrelevant here. Many other sentences are also grammatically correct for our computer, even if we may not see the point of writing them, for example, $((a \in A) \cdot (a \in B) \Rightarrow (a \in A \cap C))$. Where does this set C that has replaced B come from? We don't know, but for some specific C the above sentence might be a true statement. By defining a grammar we have in fact defined a language (in this case a language to speak about sets); in it we can state propositions, many of which are formal. From the point of view of logic, we have also defined a new universe of discourse.

Writing abstract propositions from a set of symbols that are combined according to a given grammar may be compared to playing a game invented by the surrealists, called *cadavres exquis*.[2] Someone writes down an article and an adjective; another person, without having seen the previous words, adds a noun; yet a third person, ignoring the beginning of the sentence, writes a verb, and so on. The final result is a formal sentence, grammatically correct but a priori devoid of sense, such as "The translucent peacocks sell their souls in the rain of Tunisia," and only susceptible, sometimes, of a poetical interpretation. The probability of constructing in this way the sentence "Two perpendicular lines always intersect" is rather slim, even by reducing the permissible vocabulary. We have a language, but not yet meaning, that is, a standard of truth.

In the beginning, our computer possessed the rudiments of a language to speak about concrete elements and sets, and of some known relations among them. But this language was insufficient to express abstract propositions. To form these required additional memory units and a new, larger language that extended the basic one. This extended language, which includes everything the previous one could express and more, is called a metalanguage (with respect to the former).

SOME REMARKS REGARDING TRUTH

The game of *cadavres exquis* would be boring if, from time to time, some suggestive sentences did not happen to be formed. The

[2] The expression "cadavres exquis," meaning "exquisite (or delightful) corpses," occurred during one of the first games. The players found it so poetic that they adopted it as the game's name.

mathematicians' more serious game would be even more tedious if it could never generate that precious pearl: a truth. But what is truth, in formal mathematics?

Let us take, for instance, the proposition $a \in A$. There are two possible cases: either the element a and the set A are actually present in the computer's memory, or else one of them (or both) are still mere names with no specific content. In the first case it is possible for the computer to check whether a actually occurs among the elements of A, and so decide whether the proposition is true or false. In the second case, the truth (or falsity) of the proposition may follow from some prior deduction, or its truth may be simply assumed as a hypothesis. The first case is straightforward, but the second merits further consideration.

Mathematical truth is then a consequence of the *axiomatic method*, which proceeds in four steps. To begin with, there is a universe of discourse resulting from grammatically correct propositions. These are formed, as we have already explained, from symbols representing objects (elements, sets, . . .), relations ($=$, $\in$, $\subset$, . . .) and operations ($\cap$, $\cup$, . . .). *A propositional calculus** permits one to combine the correct propositions according to the rules of logic. These rules, very similar to those introduced by Boole, can easily be taught to the computer. Basically, it is a question of specifying how to handle the operations "and," "or," "not," $=$, "if . . . , then . . . ." Actually, if we now designate the grammatically correct propositions (already stored in the *Propositions* memory unit) by the letters $a, b, c, \ldots$, then a subroutine called *Logic* (stored in some other memory unit) would be able to form the new propositions "a and b," "if a, then b," and so on, all of which would also be stored in the *Propositions* unit. This completes the second step of the method. For the benefit of the reader who would be reluctant to leave the logic up to the computer, let us point out that the logic itself could be formalized and axiomatized, thanks again to the ground-breaking work of Frege and Peano.

The third step of the axiomatic method consists in assigning to each proposition one of two "truth values." These are also represented by the symbols 0 (which we now interpret as false) and 1 (true) and stored in the memory unit labeled *Truth Values*. All concrete propositions directly verifiable by the computer can receive their truth values right away (for example, "$3 \in \{1, 2, 3\}$"—which is clearly true). Other propositions retain an indeterminate status, especially if they contain abstract concepts ("$3 \in A$," for instance), and their truth value can only be considered as an abstract "truth-value name" V, a sort of unknown of the type used in algebra, except that V can only take on one of the two values 1 or 0, true or false.

The truth value of a proposition resulting from the application of logical operations is determined by the truth value of its component propositions. For example, if both propositions a and b are assumed to be true, then the composite proposition "a and b" is true as well. This kind of rule (well known since Chryssipus) can be incorporated into the *Logic* subroutine. Hence, a sort of cloud of possible truth values hangs over all possible propositions in the formal language.

The last step gives the axiomatic method its name. It involves choosing a certain number of propositions and peremptorily deciding that they are true. These are the *axioms*,* the truth of which is decreed up front. As an example, we shall mention two of the twenty-odd axioms of elementary set theory: "If $a \in A$ and $A \subset B$, then $a \in B$," and "If $A \subset B$ and $B \subset C$, then $A \subset C$."

To help the reader grasp the general idea of the method we shall resort to an image. Let us forget the finiteness of our computer (which is here only a rhetorical device) and imagine the universe of discourse of formal set theory (or of any other mathematical theory) as a vast field sprinkled with countless trees representing all the conceivable propositions. Some of these trees, the axioms, are the source of the water of truth. The rules of logic then determine a network of innumerable channels to carry the water from tree to tree.

Since the propositional calculus generates a myriad of propositions, and since the truth value of new propositions can be deduced from that of old ones, truth will flow from the source (that is, from the axioms) to irrigate progressively the whole propositional field. A proposition whose truth is established in this way is said to be a *theorem*. Among these we find the familiar theorems deserving this designation, but also a great many other propositions that are totally irrelevant or of no interest whatsoever. All of them are nevertheless true. We can verify this by looking at the chain of logical deductions, a channel transporting the truth from its axiomatic source up to the theorem. Such a path, along which truth travels, is called a *proof*.

The axiomatic method is often daunting due to its extreme, almost disdainful, level of abstraction. Many deplore its total lack of respect for intuition, while others believe, on the contrary, that its purity serves as a protection against dangerous assumptions or misleading interpretations. Besides, the experienced mathematician can always use intuition to select truly interesting axioms, and also put his or her intelligence to work in the search for proofs. To be sure, if the axioms were chosen at random among the forest of propositions, the result would almost certainly lead nowhere—perhaps to just a few, trivial theorems—or, even worse, to contradictions, as when three of the axioms entail that the fourth one is false. The eternal miracle of mathematics makes it possible for certain axioms to generate new truths endlessly, some of which are prodigiously beautiful and subtle; the same axioms, used over and over again, and yet always equally fruitful.

Going back to our image of truth originating at the axioms and spreading through the field of propositions, it is conceivable that some of these propositions could not be reached by the flow of truth, and so are condemned to die, useless and insignificant, for lack of nutrient. They may, however, revive, if a suitable new axiom is added. It is also possible that some propositions should be accessible only through an infinite network of channels that no human can follow to its final destination, and it is at this point that there comes in the famous Gödel's theorem, which we shall discuss shortly.

Notice that each proposition possesses an opposite, which negates it, the truth of one of them obviously implying the falsehood of the other. A system of axioms is said to be contradictory if the truth of some proposition and its negation can be deduced from it. Behind an innocuous façade, a given system of axioms may conceal a deeply buried contradiction. This question of the *consistency* of axiom systems was a major source of preoccupation for mathematicians. The consistency of certain particularly simple systems—such as the elementary theory of finite sets or the arithmetic of finite numbers—has been established. For the really important and useful theories, though, the consistency question has yet to be settled, making some mathematicians uneasy.

Taming Infinity

Infinity pervades mathematics. In the case of a derivative or an integral, it is part of the process itself, the infinitely many steps leading closer and closer to the desired quantity; and when dealing with a remarkable number such as π, infinity appears in the endless sequence of digits required to write its exact decimal expression. Infinity is here, there, and everywhere. But how to tame it?

Its favorite dwelling-place is almost in front of our eyes, whenever we count: one, two, three, et cetera. What do you mean, how far is et cetera? Infinitely far. Strange notion, at the same time natural and elusive. It appears for the first time during the pre-Socratic era in Anaximander's *apeiron*, an infinite primal substance, eternal and indestructible and from which all things came. Philosophy appropriates infinity and will never cease to dream of it. Plotinus (A.D. 204–270), the founder of Neoplatonism, covers it with a mystic chasuble: the divinity is infinite in all respects, kindness, wisdom, power; it is infinity itself. This idea will thrive among most theologians, who were the first to reason about infinite qualities and infinity itself. Most of the time, they would end up face to face with a stunning paradox.

Mathematicians did not go quite as far, and, after Archimedes showed that a greatest integer cannot exist (if it did, adding 1 to it would produce an even greater number), they seemed content with accepting the dictates of intuition. Infinity again became an issue with the introduction of differential and integral calculus, in particular, the infinitesimal length of a segment that becomes arbitrarily small. They did not manage to make much progress in this direction, though, and it was only in the nineteenth century, when they could no longer escape from it, that they decided to take the bull by the horns.

There is a particular kind of infinity that seems to pave the way for all the others: the infinity of the natural numbers. We shall therefore discuss it first, resorting once again to our computer friend, to avoid being tricked by the confusion of our own mind.

The computer knows how to count: one, two, three, and so forth, up to the largest number that its memory can store in symbolic form. For the machine, there is no number beyond that one; to "think farther" it will be necessary to turn to the axiomatic method and to a more sophisticated metalanguage.

Suppose that certain memory regions are used by the computer to handle (write, store, etc.) natural numbers. These are the numbers we can write explicitly with the ten digits, for instance, 0, 1, 2, 803,712, 13. . . . The computer can always add two natural numbers, provided the size of the sum does not exceed its memory capacity. Stored in other memory units there are the signs and operation symbols of set theory, the rules of logic, instructions on how to combine propositions and how to find their truth values, and so forth. Speaking "between ourselves," all this means that we tackle the theory of natural numbers only after having developed logic and elementary set theory, and after we know what is a mathematical proposition and the difference between an axiom and a theorem.

A new memory unit will be needed to contain the (symbolic) names of abstract numbers. We shall designate them with letters such as n, p, q, and so on. It will always be possible to identify an abstract number with a concrete one and say, for instance, that $n = 13$. This identification may be a hypothesis (that is, decided by us) or the final result of some argument or computation. For instance, if we set $n = 6 + 7$, then it will follow that $n = 13$. With the help of the signs "+" and "=," the computer can write propositions about (abstract or concrete) numbers, such as $n = p + q$.

This new construction is embedded into elementary set theory by agreeing that the natural numbers are the elements of a set denoted by N. Despite its abstract character, this set is perfectly well defined thanks to certain axioms formulated by Dedekind, Frege, and Peano. Here they are:

1. 0 and 1 are natural numbers.

2. If n is any natural number, then there exists another natural number called the successor of n , which can be written as $n + 1$.

3. For every natural number n, we have $n + 1 \neq 0$ (that is, 0 is not a successor).

4. If p and q are natural numbers and $p + 1 = q + 1$, then $p = q$ (or two numbers having the same successor must be equal).

5. Let S be a subset of N having the following two properties: (i) 0 belongs to S; (ii) if a natural number p belongs to S, then $p + 1$

does too. Under these conditions, S coincides with the set N of all natural numbers.

The second axiom is named after Archimedes. It is this axiom that generates the endless sequence of natural numbers. The last axiom, due to Peano, is the basis of the induction principle, which plays a fundamental role in many mathematical proofs. It would be instructive to give a simple example of its application, even if it entails a short digression. It is said that while Gauss was attending elementary school, his teacher had once given to the class the following exercise: add 2 to 1, then add 3 to the previous sum, and continue like this until you reach 100. The teacher expected that while the students were busy adding all those numbers, he could enjoy a peaceful break, long enough to digest his meal. But after only a few minutes, he noticed that Gauss had stopped calculating. Intrigued, he went to check the child's copybook and found that, after a few additions, Gauss had multiplied 100 by 101 and then divided the product by 2, obtaining 5,050, which is the right answer. If he had relied on axiom 5, Gauss might have remarked that $1 + 2 = 3$, $1 + 2 + 3 = 6$, $1 + 2 + 3 + 4 = 10$, and that if the last number added is n, then the sum equals $n(n + 1)/2$. Hence his simple calculation of the correct answer.[3]

There remains the question of justifying why the above formula is valid for every natural number n. The induction principle comes to the rescue. Call S the set of natural numbers n for which the addition formula $1 + 2 + \cdots + n = n(n + 1)/2$ is true. It is easy to check that on adding $n + 1$ to each side of the equation, the right-hand side becomes $(n + 1)(n + 2)/2$. Hence, if n belongs to S, so does $n + 1$. But the number $n = 0$ also belongs to S, for $0 = 0(0 + 1)/2$. Hence, according to axiom 5, S coincides with N, in other words, the formula is universally valid.

Peano's system completely characterizes the set of natural numbers with just a few, perfectly clear axioms that are easy to apply in practice. We may wonder at its simplicity, but equally at the fact that it took more than two thousand years to discover.

The axiomatic method permits the systematic construction of all numbers. First the integers, or whole numbers (the non-negative

[3] As a matter of fact, young Gauss added 1 to 100 and found 101, and the same result for 2 + 99, and so on. Then he only had to multiply 101 by the number of such partial sums, namely, 50.

integers are precisely the natural numbers); then the fractions (also called rational numbers, because they appear as ratios of integers), both positive and negative; next the so-called real numbers, which can be written in decimal form (with, perhaps, infinitely many decimals), followed by the complex numbers. All these constructions are based on a method that remains clear and rigorous at each stage. The same axiomatic approach may be applied to other important notions, such as that of a group, to various geometries, and to the whole arsenal of analysis. Different types of infinity, far more fearsome than the infinity of the natural numbers, may cross our path, but there again, the net of axioms and logic will be able to support them without breaking. We shall not engage any further down this road, though, for it leads to the immensity of modern mathematics.

Today's Mathematics

Present-day mathematics is entirely based on the axiomatic approach, at the origin of which there is a system of symbols with no direct connection to reality and governed by its own rules. The principal characteristic of this mathematics is its total submission to logic, which is also formal and symbolic. This level of abstraction does not imply that there is no more room for imagination. On the contrary, the choice of relevant axioms, the conjecturing of interesting or far-reaching theorems, the search for (or the refinement of) proofs, and the discovery of the inspiring analogies that pervade the vast domain of mathematics, none of all that would be possible without the creative power of imagination.

This imagination does not spare any effort in its pursuit of novelty, and, if symbols remain the foundation, no commandment enjoins the mathematician to stay within their bounds. In fact, rather than use symbols, mathematicians prefer by far to speak a language as similar as possible to ordinary language. They do not despise suggestive images either, and the words they employ are often a far cry from the arid strings of abstruse symbols processed by computing machines. They say "sets," "spaces," "numbers," "neighborhood," "ideals," "metric," "curvature," "filters," "choice," "applications," "groups," "intersection," "union," "distribution," each word referring to a concept based on a strict axiom system, but no less evocative for that reason. Other words may be more intimidating, "isomorphism," "functor," "topology," but a little etymology is often enough to clarify them. All in all, the language of mathematics is less hermetic than medical terminology and rather comparable to the language of botany. The difficulty lies less in learning it than in speaking it fluently.

Contemporary mathematics is tremendously rich, and its complete exploration would require a lifetime. To draw up an atlas of it is therefore out of the question, especially since the continents have changed places—as well as shape—since the classical period. Words like arithmetic (or number theory), algebra, geometry, and analysis no longer mean what they used to, due to the increase in knowledge and in the variety of the topics covered. A number of different "structures" stand out, as in a relief map: collective properties (sets), properties of closeness (topology), of operations (rings, groups, . . .), of functions (a concept having multiple forms and everywhere present), each of these structures branching out into many other substructures. This new cartography corresponds to particular ways of grouping the axioms, thus revealing unexpected connections between seemingly unrelated domains of application.

The choice of the axioms is not arbitrary. If someone wishes to introduce a new axiom without a serious investigation, he or she will most likely end up finding a heap of worthless stones instead of the vein of gold hoped for. The choice of this or that axiom, to simplify a theory or promising to blossom into a new one, can only be the result of a study of hundreds of relevant examples, or of a penetrating intuition, as much the fruit of effort as the product of intelligence. Only if its consequences are rich in new results and in solutions to open questions will a proposed new axiom or concept be adopted.

Mathematical ideas are like living species competing with each other: in order to survive, they must be useful, well adapted, and, above all, fertile. This much should be obvious, for otherwise they would go unnoticed. What is less obvious, actually extraordinary, is the proliferation of such fertility, the vastness of the ground it has covered. No one can explain such a phenomenon in a satisfactory manner, even if it is one of the most amazing facts in the history of ideas. What's more, there do not appear to be any limits to the potential of mathematics.

But we have already spoken enough about mathematics. Let us now complete our brief outline of its history. The period of reorganization through the axiomatic approach, with the inevitable false starts that accompany any enterprise of such a magnitude, lasts approximately one hundred years, from 1850 to 1950. Several names stand out: Weierstrass, Dedekind, Cantor, Frege, Peano, Hilbert, Russell, and Whitehead. Many others also took part in the effort, but we unfortunately cannot go into the details of their individual contributions. In a certain way, most of the program was explicitly worked out by the multifaced Nicolas Bourbaki (a pseudonym adopted by a team of mathematicians whose composition continues to change).

Since the middle of the twentieth century though, radical axiomatics has ceased to occupy center stage. From the vantage point of hindsight, some revisions were even made, and axiomatic systems that were too general to be fruitful became obsolete. A fair balance was finally struck, and today we are witnessing a new harvest, with a preponderance of results over work on the foundations.

The Crisis in the Foundations of Set Theory

We would be presenting a distorted picture of history if we gave the impression that a sweeping transformation such as the one just described took place without drawbacks or opposition. Two famous episodes of this saga deserve to be mentioned, if only to size up the significance of the changes that took place: the crisis in the foundations of set theory, in 1902, and the discovery of Gödel's incompleteness theorem during the years 1930–1931.

What is known as the crisis in set theory is a striking event that deserves to be replayed in the spirit of what it was, that is, a drama lacking neither heartbreaking nor noble undertones. Here it is then, more or less as it might be seen on the stage.

Two characters are present as the curtain rises, Gottlob Frege and Bertrand Russell. The action is set in a temple, that of the goddesses Mathematics and Logic. At the back are full-size portraits of the great priests of the time: David Hilbert and Henri Poincaré. Other pictures, in subdued tones, depict Dedekind, Peano, and Cantor. A portrait of Frege himself appears on an easel in the foreground; it has just been retrieved from the storage room after a long stay there.

The actor playing Frege appears to be in his fifties. He is unassuming but betrays a unique passion that can only be inspired by truth. Almost twenty-five years have elapsed since the publication of his short book on logic which had initially gone unnoticed. Bertrand Russell is thirty years old. He has the unmistakably sharp traits of an aristocrat and speaks with a slight Cambridge accent.

FREGE: Yes, my final book on the theory of sets is due to appear soon. It took me twenty long years of hard work, but it was perhaps worth the effort.

RUSSELL: You know very well what I think of it. Nothing as important as your first book had been written in logic since Aristotle; and your latest one, I believe, should definitely establish mathematics on a solid base. What an achievement for the honor of the human mind![4]

FREGE: Let us not exaggerate. It is true, at any rate, that the logic is sufficiently clear. As for the mathematics, I think one should begin with set theory and build everything on it. In fact, there is nothing simpler or more transparent than a set. When you speak of a collection of objects, everybody knows what you are talking about.

RUSSELL: Yes, it appears to be quite obvious, and yet, I have one nagging reservation.

FREGE: Which one?

RUSSELL: Something in your *Begriffschrift* that puzzles me. You say there, essentially, that an arbitrary set, and I insist on the term "arbitrary," may always be taken as an element of another set. Do you still think so?

FREGE: More than ever. A major part of my new book is based on that fact, and the idea is repeatedly exploited. Do you have an objection? I thought it to be obvious. What's wrong with the idea that any object can always be included in a set along with other objects?

RUSSELL: That is certainly what our intuition tells us. But I wonder if we can always trust it, and if it is not possible that intuition may deceive us when left unchecked even for an instant.

FREGE: All right, I can see that you have found a skeleton in the closet. Better take it out. What is it?

---

[4] This expression is due to Hilbert.

RUSSELL: Do you agree that, in principle, certain sets may contain themselves as elements?

FREGE: It is at any rate a direct consequence of what we said earlier. If you asked me for an example, I would propose the catalog of a library, which can be one of the books placed on a shelf of the same library; or the word "dictionary" in a dictionary; or God, who says "I am who I am"; or the table of contents of a book, which contains the table of contents, or even . . .

RUSSELL: I see. But let us consider all the others instead, and designate by *A* the set of all those sets that are not elements of themselves. Now let me ask you a question: Does this set *A* belong to itself?

FREGE: Let's see, this should not be difficult. Suppose it does, that is, that *A* belongs to *A*. Now, by definition, the elements of *A* are those sets that do not belong to themselves. Thus, assuming that the answer to your question is "yes," we have a contradiction. Therefore, the answer must be "no."

RUSSELL: Are you sure?

FREGE: If I answer "no," this implies that *A* does not belong to *A*. But then, by the very definition of *A*, it follows that *A does* belong to *A*. Good Lord, you are absolutely right! No matter which path we choose, it leads to a contradiction. This is a paradox, what am I saying? An aporia, a catastrophe! It is the principle of the excluded middle that you have just called into question. But this is impossible, we cannot reject this principle, for there would be no logic left, all thought would collapse.

RUSSELL: I can see only one way out: to repeal what you have said in the past and start all over from the beginning.

FREGE, *after a moment's reflection*: There is no other solution. Naturally, my great project of rebuilding mathematics is shattered to pieces. Just when I thought I had succeeded! But, you know, what you have found is truly amazing, extraordinary. Congratulations! It is a while since I came across something so interesting! (*He leaves walking unsteadily, smiling and talking to himself.*)

RUSSELL, *watching Frege leave*: What a demonstration of intellectual integrity! Such grace! I have never seen anyone pursue truth as honestly as he does. He was about to culminate at last a lifelong endeavor, he who had been so often passed over in favor of others who did not deserve it. . . . He did not care, and when told that one of his most fundamental hypotheses is wrong, how does he react? His intellectual pleasure overwhelms his personal disappointment. It's almost superhuman. What an interior strength a man can summon if he devotes himself entirely to knowledge and creation, rather than to a vain search for honors and celebrity! What a lesson![5] (*He also leaves.*)

THE CHOIR: The temple has been shaken and it is cracking. Is it an earthquake? Paradoxes are piling up. The Cretan liar has been resuscitated. There are also Richard's and Burali-Forti's paradoxes, besides Russell's. Are we to become everybody's laughing-stock when it has been pointed out that an eleven-word sentence suffices to define "the smallest number impossible to name with less than twelve words"? Is logic only an illusion?

HILBERT, *entering the room*: Calm down, please, and do not panic. Look those fearsome paradoxes straight in the eye. They are all alike. They all carry the same sign, that of the whole considered as a part. The library's catalog is a list of *all* books. Epimenides, the Cretan, says that *all* Cretans are liars. Your eleven-word sentence refers to *all* possible definitions of a number. This story shows only one thing: that Frege had not gone far enough in his efforts to formalize mathematics. He thought he could trust his intuition, if only a little, regarding sets, which appear to be so limpid. It was his sole mistake, and it is our duty to correct it. From now on, logic and mathematics will be entirely formal. (*He leaves, followed by a thoughtful Zermelo, who would take up the task proclaimed by Hilbert.*)

GÖDEL'S INCOMPLETENESS THEOREM

It does not happen often that an event concerning mathematics reaches (and stuns) the external world. But this is precisely what happened in the nineteen-thirties with a certain theorem of Gödel, considered to have dealt the human mind a humiliating blow.

What was it all about? Kurt Gödel was a disciple of David Hilbert, and he was working on one of the master's great projects: to demonstrate the consistency of the axioms of arithmetic and so establish, once and for all, that at least this branch of mathematics

---

[5] These are almost the exact words employed by Russell in a letter to Jean Van Heijenoort, where he talks about Frege.

is forever immune to internal contradictions. Paraphrasing the Romans, Hilbert wanted to be able to say, "I build here a monument for all eternity."

Hilbert's axiomatic system for arithmetic was formulated in terms of symbols and signs (as we have indicated earlier) that included the usual operations: addition, subtraction (whenever the result is a natural number), multiplication, division with remainder, and exponentiation. The necessary axioms captured the basic properties of the natural numbers and of these operations. Hilbert then considered the set of all propositions (about the natural numbers) that can be expressed in the formal language. The problem was to demonstrate that every such proposition had, at least in principle, a truth value resulting from a proof—that is, from a finite chain of logical implications having its source in the axioms.

Gödel showed that one can actually assign a truth value to certain propositions without going through a (formal) proof, but only with the help of a higher-level theory equipped with a metalanguage (a concept we have discussed earlier). He did answer Hilbert's question, although not in the way the latter expected.

Indeed, Gödel's feat was to show that there exist propositions that are true (from the point of view of the metalanguage) but whose truth is impossible to establish by a (formal) proof of finite length (hence, the axiomatization of arithmetic is "incomplete"). And so, if a mathematician only accepts as true what can be logically proved from the axioms, there will be some propositions that will remain (for him or her) forever undecidable, for they can neither be proved nor disproved.

Gödel's result may be easily understood if we employ once again the analogy of truth flowing from the source (the axioms) toward the trees (the propositions). Then, Gödel tells us that the whole forest of trees cannot be irrigated by a network of channels of finite length, and that to reach certain propositions would require an infinitely long path. There is nothing, after all, really surprising in this result, much less a reason to claim that the human mind has taken a beating. We must accept that there exist unsolvable problems, probably many. How many of these problems have actually been found? Only very few. Why is it surprising that every proposition is certainly true or false, though the slow process of proof cannot always decide which? Wouldn't the contrary have seemed unnatural? What is more serious is the uncertainty regarding the consistency of arithmetic. We simply do not know that no contradiction lurks in the shadow of the axioms. But isn't this the price to pay for their bountiful consequences? Once again, an almost insane expectation of human ambition, the dream of building for eternity, met its nemesis. This is as it should be.

All things considered, Gödel's incompleteness theorem is indeed an exploit of human intelligence. But at the same time it establishes its limits, without, however, destroying or spoiling its achievements. It only reminds intelligence that it too is perhaps mortal.

A Tentative Conclusion

It now appears certain that mathematics is strictly a science of relations, uncommitted as to the specific objects these relations are about. Though mathematics is used as a tool, a language, and a logical framework in some physical sciences and notably in physics, it does not have any meaning by itself. When seen through mathematical goggles, every physical science provides its own metalanguage that comes with a particular meaning. One can express the same thing more mundanely by saying that mathematics cannot help us find the real meaning in a formal science; the meaning has to be found in that science itself. This lesson will be essential when we face the cliffs of quantum mechanics.

The Philosophy of Mathematics

W E SHALL NOW EXAMINE the question of the meaning of mathematics. In particular, we shall listen to what philosophers and mathematicians inclined to philosophize think of this science. Whether their answers should be endorsed or rejected remains to be seen, but the present situation will prove sufficiently complex to at least allow us to estimate the stretch of road yet to be traveled.

WHAT IS MATHEMATICS?

What is mathematics, this strange outgrowth of reason, where does it come from, and what is its nature? This question is as old as the subject itself, but it usually attracts only a small number of philosophers and mathematicians. Who else would care? No one. However, given the formal character of many explanations of nature and its mysteries, we should be well advised to take a closer look. What if mathematics should conceal one of the keys to knowledge, if the "No one but a geometer may enter here" meant, unexpectedly, that the royal path to philosophy begins with the above question? Who would then be willing to ignore it?

Mathematics has often been conceived as dwelling in a divine world, filled with a perfect light. This is Plato's view, as well as Nicolas de Cusa's, among a great number of others. It was then possible to believe that the proofs given by the mathematician, those perfect models of truth attained through a safe and triumphant method, drew their strength from some divine grace; a grace to be treated with respect and successfully applied to other domains. And so, a good portion of theology at the end of antiquity and during Scholasticism, culminating with Saint Augustine and Thomas Aquinas, uses mathematics as a model and inspiration. The most striking example is offered by Spinoza. The truth of the propositions in his *Ethics* follows, or at least the author so pretends, by the force of arguments that proceed "in the manner of the geometers." A similar attitude also pervades the philosophy of Leibniz, who was, by the way, an admirer of Spinoza.

This sort of kinship suggests that the question of the nature of mathematics may sometimes get mixed up with others more appealing to a philosopher. In fact, it is difficult to dissociate it completely from questions regarding the character and the power of reason, or the existence of order in nature. The least we can say is that the question is more important than may initially appear, and certainly more difficult.

Trying to define mathematics is hopeless, for it would be tantamount to either knowing already what it is, or capturing only its external features. Etymology is of no use either, for it simply evokes things that are well known and well understood, or perhaps a knowledge for initiates, when it becomes a *mathesis*. Therefore, we shall start by noting some of mathematics' most striking characteristics. These will serve to evaluate the various philosophical theories, because many of them fail to take into consideration one or the other of these characteristics, while a satisfactory philosophy of mathematics should account for all of them.

The first attribute, its beauty, has often been put forward. It is a rather peculiar beauty, known only to those who come sufficiently close to it (but isn't this true of any beauty?), difficult to describe in a language other than its own, and impervious to the language of poetry. True, it is sometimes embodied in the harmony of proportions or in the elegant forms of a work of art. According to Bertrand Russell, mathematics possesses not only truth but also supreme beauty, an austere and cold beauty comparable to that of a sculpture. True beatitude, exaltation, and the feeling of being more than human may be experienced in mathematics as certainly as they may be found in poetry. Plotinus, many years earlier, had gone further, and even turned the interpretation around. The beauty of a statue, of Zeus, for instance, attains perfection when the artist succeeds in translating into marble something of the essence, of the Form, of the god—a present-day Neoplatonist would perhaps say, in the same spirit, that Velasquez or Monet have each captured something of the essence of light. For Plotinus, the beauty of mathematics or, better still, of a coherent philosophy, offered the best possible model for casting a piece of art.

Another important feature of mathematics is its fertility, the character we have tried to emphasize in the previous chapter. But is "fertility" the appropriate term? Profusion might be a better word to describe the overabundance of this torrent wide as a sea, comparable to the birth of the world as related in the *Mahâbhârata*. I have chosen this particular analogy on purpose because mathematics has one hundred arms and one thousand breasts, and it is infuriating to see it reduced by the shortsighted to the size of a poor skinny thing. This enduring fertility is clearly an essential trait of its nature, one which should arouse a desire to find out what makes it possible.

Three other distinctive characteristics of mathematics have been mentioned earlier. The first one is its close relationship with logic, so much so that it would be impossible to tell where one ends and the other begins. Second is this possibility of reducing mathematics to a mere manipulation of symbols, by which it detaches itself completely from any concrete reality. Actually, nothing is more removed from the real world, even though it was reality that provided mathematics with its initial motivation. It is abstraction in its most extreme form (to abstract: to pull out, to unroot) and, since the birth of Greek geometry, this abstraction unroots mathematics from physical reality. And yet, despite all that, the third character of mathematics is the intimate connection that it maintains with reality, in the sense that the physical sciences—and physics in particular—cannot do without its language or its concepts. Let us mention one last feature, not to be forgotten even if entirely prosaic: mathematics is a product of the human brain, and it was created by humans who live in a society.

If we leave out, regretfully, its beauty—which cannot be captured by words—the distinguishing characteristics of mathematics are its fertility, the possibility of being reduced to symbols, a high correspondence with reality, and the fact of being the result of a human activity. Within this framework, it is convenient to classify the philosophies of mathematics into two categories: ontological and sociological. The first one is concerned with its intrinsic qualities, what is mathematics considered by itself; the second category tries to understand mathematics within a human setting, the conditions under which humans have built the great edifice of mathematics.

MATHEMATICAL REALISM

The most ancient philosophical theory, still very much alive today, maintains that there exists a world, different from concrete reality, where mathematical truths properly belong. Plato was the first to propose it, and in his formulation this other world is the universe of Ideas. This point of view, whose origin goes back to Pythagoras, is usually known as Platonism. Others prefer the name mathematical *realism*,* to stress the radical assumption of the existence of a separate reality. We shall employ this second designation, which has the advantage of avoiding all ambiguity.

Mathematical realism inspired a passage from Descartes' Fifth Meditation that we have already quoted but is worth repeating: "When I imagine a triangle, even if perhaps such a figure is nowhere in the world to be found except in my own mind, and it has never been, it does nevertheless exhibit a certain nature, or form, or definite essence of this figure, which is immutable and eternal, and which I have not created, and which does not depend upon my mind in any way whatsoever; as appears to be the case since one can demonstrate certain properties of this triangle."

Hermite says basically the same thing in a letter to Stieljes: "I believe that the numbers and functions of analysis are not arbitrary creations of our mind. I think that they exist independently of us with the same kind of necessity as the things of objective reality, and that we find them, or discover them, in the same way physicists, chemists, or zoologists do."

Bertrand Russell, certainly not a neophyte in philosophical matters, expressed the same idea when he wrote[1] that the number 2 must be an entity possessing an ontological reality even if it is not in any mind. For him, cognition is necessarily recognition, since otherwise it would be nothing but illusion. He believed that arithmetic must be discovered just as Columbus discovered the West Indies, and that we no more create the numbers than he created the Indians. The number 2 is not merely a purely mental creation, but an entity that can be the object of thought. According to Russell, anything that can be thought possesses an ontological reality,

---

[1] In *The Principles of Mathematics*.

which is a precondition for that thought, and not its result. As for the existence of the objects of thought, nothing may be concluded from the fact that they thought, for they are certainly not in the mind thinking them. In conclusion, Russell maintains that the objects of our representation as such do not possess any special kind of reality.

For Jean Dieudonné,[2] it is certainly "quite difficult to describe the ideas mathematicians have, which vary from one person to another." He adds, however, "They generally acknowledge that mathematical objects possess a 'reality' distinct from sensory reality, perhaps similar to the reality that Plato claimed for his Ideas."

As for Alain Connes,[3] he confesses, "I think that my position is not far from the realistic point of view. For me, the sequence of prime numbers, for instance, has a reality that is more permanent than the material reality surrounding us. . . . The axiomatic method, to mention only that one, allows the mathematician to venture beyond the familiar territories. . . . Mathematical reality possesses a consistency truly superior to that of sensory intuition, an unexplained coherence that is independent of our reasoning system."

There is plenty of evidence supporting the realists' point of view: nearly all of them are creative mathematicians, and they know well the familiar feeling of discovery, constantly renewed, they are talking about. It is also what Connes calls coherence that distinguishes the vision of the mathematician from any other art form: mathematics in its entirety, from Pythagoras' theorem to the proof of the most recent result, possesses an almost complete unity, it is more like a single piece of work than a myriad of parts put together. This coherence can also take the form of a perfect harmony between the question that one poses and the answer one expects to find; or that of a generalization that transforms an innocuous theorem into a powerful theory; or even the presence everywhere in the edifice of analogies that repeat and reinforce themselves in one thousand different ways. Coherence also manifests itself when, from a multitude of seemingly arbitrary inventions, new structures suddenly take form, structures that prove amazingly fertile: distributions,

[2] In the introduction to his *Abrégé d'histoire des mathématiques* (Paris: Hermann, 1978).

[3] See Jean-Pierre Changeux and Alain Connes, *Matière à pensée* (Paris: Odile Jacob, 1989).

metric spaces, Hilbert or Banach spaces. These examples also show, unfortunately, the difficulty for mathematicians in communicating what they see: only they can really know the wonders that delight them so much; others, the noninitiated, must make do with the pale and simplified image glimpsed from the mathematician's description.

The realists never doubt that they are advancing on solid ground that has always existed, and which they are merely exploring. Their approach resembles in some ways the exploration of a virgin forest, wild and thick. Occasionally, the explorers may run into a large clearing in this jungle, but progress is usually made only through demonstration, in a slow, almost creeping fashion, demanding a safe footing for each step. The mathematician at work advances with eyes fixed on the ground, knowing that some supreme truths will only be reached after a long journey, and that from the top of these truths the view, magnificent, extends all the way to the horizon. Spinoza so imagined a knowledge of the "third kind," transcending that acquired through a patient proof, which is only of the "second kind."

All these men and women through the ages keep reporting their impressions and the things they have observed: the existence of a vast continent that each of them has partially explored; the joy of having visited some bushes of truth; the solidity of the territory and of its interconnections—coherence. . . . Like mystics, they speak of other horizons, where they have contemplated oceans of light. Doubt and suspicion may then assail the listener: Have they not merely been fooled by mirages, tricked by a brain too proud of its own power, or possessed by a dream so strong that it appears more real than reality? These are the reasons why mathematical realism is so often dismissed by the skeptics as a mild form of illuminism.

Objections of this kind can only be answered with the help of history, the only objective factor in things concerning the mind. It is interesting to listen again to Alain Connes, when he maintains that the coherence perceived by mathematicians is independent of any particular reasoning mechanism. To put it more vividly, we may say that mathematicians all agree in recognizing the same features of the explored territory, independently of the paths they have followed.

As an example, we shall go back to the history of analysis between the seventeenth and the nineteenth centuries. At the time, a large number of truly amazing results was available, but the frailty of their foundations had rendered many mathematicians uneasy. A vast program of revision and criticism was then undertaken, unparalleled in the history of ideas. The violent attacks brought against theology after the end of Scholasticism had been innocent taps on the wrist compared to the fury that mathematicians displayed against their own dwelling—the whole building should have collapsed, leaving but the desolation of scattered ruins. What happened instead? The edifice gained in strength and majesty, rising even higher and more spacious than before, the old cracks sealed, the weaknesses repaired. And yet, in a certain sense, everything had changed, axiomatics had replaced intuition, and the structure now reflected a novel order; the methods of reasoning had changed, but the consistency of the results, old and new, came out enhanced and better secured.

Mathematicians constantly run into this kind of lesson in the course of their research work. They get the impression that what they find is not necessarily what they expected, but rather what the force of circumstances imposes, with all the necessity of a world that exists on its own.

The objections against mathematical realism are primarily due to preconceived ideas about the nature of reality. But these preconceptions appear less convincing when confronted with the image of reality conveyed by modern physics. I shall restrict myself to one example. The philosopher André Darbon published a book devoted to Russell's "logistic" theories, full of valuable ideas and information.[4] Unfortunately, in the last chapter he lashes out against Russell's realism, calling it "childish," while at the same time supporting his argument with well-thought and learned considerations. Among these, Darbon takes for granted the impossibility in principle of indiscernibles—and in this he has Leibniz on his side—but his position had been refuted twenty years earlier by experimental physics. This example reveals a recurrent element in the criticism of realism: the presence of preconceived ontologies, which are no more believable than those they seek to reject. We shall not elaborate this topic any further.

More recent criticism of a different type deserves to be men-

[4] André Darbon, *La Philosophie des mathématiques* (Paris: Presses universitaires de France, 1949).

tioned. In his book *Matière à pensée*, written jointly with Alain Connes in dialogue form, Jean-Pierre Changeux presents the point of view of a naturalist and specialist of the brain. He notes that brain structures for perception and for the organization of functions exhibit an internal predisposition, conscious or unconscious, to handle symbols. Our brain invents symbols because it itself operates by processing concrete symbols: its own signals. When it marvels at the discoveries it makes through the power of thought alone, it might only be admiring itself, as the marvel of nature that it is. The coherence we encounter in the products of our brain could be the reflection of the amazing internal harmony of our thinking machine.

This objection has an even greater impact because it strikes right at the heart of what mathematicians cherish most. But it seems to me that Changeux' criticism has a flaw a naturalist should have recognized: the marvel of mathematics is reduced to another marvel, that of our own brain. This wonder of nature, like everything alive, is the result of billions of years of evolution, which should explain its near-perfection. But there is more. For what reason has evolution endowed the brain with such properties if not because they are useful for survival, which presupposes a corresponding order proper to the external world: an order immanent in reality? When science explores this order and ends up discovering the principles of physics, it bumps into mathematics once again, but this time as a necessity associated with reality, and no longer as a product of the independent activity of the brain. Summing up, we have been running around in circles, and the answer proposed by Changeux is not really an answer; it is, at best, only one of the projectors that help to illuminate part of the picture.

We shall have to come back to this question, but we can already begin with a simple observation: no serious discussion of mathematical realism can be carried out independently of an examination of the laws of the physical world, that is, the nature of mathematics is inseparable from the nature of those laws. Their strange characters are intimately interconnected, and no one can reject in principle the existence of an intangible reality in the name of common sense when this very common sense is being attacked by physics. It would be a mistake today to build a philosophy of mathematics independently of a philosophy of the physical sciences.

Going back to the characteristics of mathematics listed at the end of the previous section—beauty, coherence, fertility, agreement with the laws of physical reality—mathematical realism accommodates them all. The only problem with realism, and not a minor one, is getting people to accept the existence of something intangible, something one cannot point at and say, "There, it is that"; the existence, in short, of a reality that could be neither immediate nor perceptible.

NOMINALISM

Every realism is inevitably confronted with a nominalism. In the present case, the most extreme form of nominalism consists in claiming that mathematics is a game played according to arbitrary rules, similar to chess, only more complicated. This position will not be found among mathematicians but rather among philosophers, such as André Darbon, whom we have quoted above.

Darbon considers mathematics to be a "hypothetico-deductive" construction, a strictly human invention based on hypotheses that can be freely chosen, including, if so desired, the rules of reasoning themselves. One plays the game by drawing conclusions from the hypotheses by means of the given rules.

Surely, the above description fits perfectly well one of the aspects of doing mathematics. However, the philosophical claim of nominalism is that mathematics reduces to such a "game," that there is nothing else to it. Although this doctrine originated with Leibniz, it would be unfair to condense the ideas of the philosopher of the monads into such a simplistic form. It is better to listen to what an expert has to say. Here is Jean Dieudonné's learned opinion on the existence and the interest of purely arbitrary constructions in mathematics: "It appears that mathematical problems of consequence are somewhat like living beings, having a 'natural' evolution that it is in our interest to respect. Artificial systems of axioms, created for the sole purpose of generalizing some familiar problems in an arbitrary way, have seldom had any remarkable consequences." In other words, a theory that assumes mathematical hypotheses to be gratuitous and arbitrary is immediately confronted with the mystery of their fecundity and applicability.

Darbon may argue that the origins of geometry and arithmetic were rooted in concrete reality. But this argument only begs the question of why reality obeys laws that only mathematics is able to express. In conclusion, strict nominalism fails on three major counts: fecundity, coherence, and correspondence between mathematics and reality.

Some nominalists, such as Darbon himself, tried to justify their position by calling attention to the results of the so-called formalist school, whose members are strict defenders of the axiomatic method. Actually, the use (or the rejection) of this method has nothing to do with the nature of mathematics, but is only a question of which method is more appropriate. Nothing prevents a realist from employing the axiomatic method, and there is no shortage of examples of mathematicians who did so. The axiomatic method only degenerates, so to speak, into nominalism, if one maintains that mathematics is exclusively a matter of method and refuses any reflection on the scope and the coherence of mathematical results. It is true that for diehard formalists mathematics is entirely reducible to symbol manipulation and, in this sense, it coincides with nominalism. However, few mathematicians are prepared to go that far.

Since we have just discussed the formalist school, we might as well mention a few other points of view that deserve attention, "logicism," for instance, which is based on the axiomatic method but focuses on the reasoning process. What counts is to construct "proofs," and the axioms of logic on which these rest are supposed to be the most fundamental. They are considered to be completely general, free from arbitrary elements and applicable to any form of rational thought. We know them intuitively or, as Hume would put it, as a result of the accumulation of facts and for their role in the structuring of language. This conception was developed mostly by Russell and Whitehead, and it is sometimes schematically described by saying that mathematics reduces to logic. This means that logic possesses an ontological basis that partly determines mathematics. Hence, mathematics is essentially realistic, but also in part arbitrary.

The so-called intuitionistic school, whose principal architect was Brouwer, would like to restrict the domain of mathematics to what the imagination can envisage, and it stands against what it considers to be the excesses of modern formalism. This leads it in practice to eliminate certain axioms, and to restrict the scope of those that remain to what intuitionists deem legitimate. We shall not insist on this controversial topic, which is rather technical. Clearly, intuitionism must be regarded as a form of mathematical realism, but of a very peculiar sort. One might describe it by saying that, for the intuitionists, there exists a mathematical reality accessible through intuition, but which is not immanent (that is, given once and for all and existing by itself); on the contrary, it is a reality that mathematicians progressively construct. Put another way: the initial reality is the universe, which gives rise to human beings, some of whom are mathematicians, and whose task is to increase the intellectual content of the universe. . . .

MATHEMATICAL SOCIOLOGISM

There exists yet another important position, variously called the scientific community theory or mathematical sociologism. According to its followers, mathematics is essentially what the community of mathematicians decides it is. Mathematical sociologism is therefore a form of nominalism, whose main interest lies in the fact that it highlights the daily activities of men and women who form a microsociety, an international association.

As with any select club, new members of the brotherhood are subjected to initiation rituals by means of tests and exams, which culminate with a thesis. Teaching is intended to disseminate and to control the doctrines that prevail at a given time. Hence, in the middle of the twentieth century the dominant credo was axiomatics, as formulated by Hilbert and codified by Bourbaki in his books. Shortly afterward, it triumphantly entered school curricula under the name modern mathematics. Since then, the development of computers has contributed to reorientating the interests of the mathematical community, and we are witnessing a shift toward combinatorics and away from other domains such as analysis, where infinity plays a central role.

Philosophers and mathematicians who have been influenced by this school, such as Raymond Wilder and René Thom, go far beyond traditional nominalism by taking into account the profusion of mathematics in all its manifestations, particularly from the perspective of the creativity factor. Creativity in mathematics has been analyzed in depth since Poincaré told how the idea of a "Fuchsian function" occurred to him as he was about to board a bus. From the intense investigations on the subject it appears that new ideas emerge in a flash, provided the researcher has been constantly focusing his or her thoughts on the problem. Regarding this "private epistemology," the mathematical community plays the role of both critic and amplifier, ultimately deciding on the fate of the new ideas: whether they deserve to be reflected upon and communicated, or abandoned and condemned to oblivion.

Some may fear that this reduction of mathematics to such a corporate agreement may transform it into a mere game of conventions and conveniences, if not of fashion, where very little would distinguish the mathematical community from a bridge club. To counterbalance such a tendency, it is important to stress the fact that mathematics is also a tool for acquiring knowledge. However, among the supporters of this position, there does not seem to exist a common view as to the reasons why mathematics is so amazingly efficient when used in the other sciences. This mystery is not one of their main preoccupations, though, except insofar as utility serves as a criterion to judge the potential interest of new research avenues. They consider that the most characteristic feature of mathematics is the notion of proof, the aspect that singles out their discipline from all other domains of knowledge.

Imre Lakatos has carried out a detailed and far-reaching analysis of the notion of proof, in both its theoretical and practical aspects. He has studied how a mathematical "truth" takes form in the private epistemology of the mathematician and in the community, and he divided this process into several stages. According to Lakatos, everything begins with a conjecture, that is, a hypothesis assuming that a certain theorem is probably true. This conjecture might have been suggested by some examples, or be an idea that has occurred to the mathematician and which, due to its beauty, deserves further examination. The mathematical literature contains a great number of conjectures, famous or otherwise, collected through the centuries. Both the sociological school and Lakatos are interested in how these conjectures come about, but this aspect belongs to the reflection on creativity already mentioned.

The proof of a given conjecture may be tackled by a single mathematician, by a whole team, or even by several groups working together or competing with one another. Whatever the case, they begin by reviewing several known methods of proof that appear promising. This is followed by an evaluation of the likelihood of success of each of them. What Lakatos calls a "proof scheme" is the decomposition of the main argument into a series of lemmas, or secondary conjectures. While some of these lemmas may be proved without much difficulty, others may remain at the hypothesis level until a satisfactory proof is found.

The proof scheme is then tested, by searching for counterexamples that would refute some lemma initially believed to be true. The discovery of a counterexample may either discredit some lemma(s), resulting in a reconsideration of the proof scheme, or demolish the whole conjecture by demonstrating that it is false. In the latter case, the search for a proof was not necessarily a total waste of time, for the intellectual effort deployed might have suggested new possibilities and lead to a new round of creativity. In some cases, the counterexamples found might appear to be so extravagant that, rather than rejecting the conjecture as false, it is preferable to restrict the hypotheses conditioning its validity (Lakatos calls such a course of action monster-barring). The search thus proceeds by stages, and it might eventually produce a satisfactory proof of the original (or modified) conjecture; or else the questioning of the hypotheses may end up, in some extreme cases, provoking a revision of the axioms of the theory on which the conjecture rests.

Lakatos claims that this is precisely the kind of process that leads to results likely to be accepted by the mathematical community. It follows that the notion of "truth" is then more a matter of convention than the attainment of something absolute and undisputable. This point of view may be illustrated with the example of analysis: between the eighteenth and the nineteenth centuries the criteria of rigor were considerably sharpened, leading to the complete revamping of the axioms we have already reported. Mathematical proofs mature with age, just like mathematics itself.

The above considerations are perfectly valid, and faithfully reflect reality; they are based on serious investigations on the way mathematics is done. We may, however, wonder whether we have learned a great deal, after all, about the nature of mathematics. A realist may object that a similar study could be made of the evolution of maps and charts between antiquity and the Renaissance. From the stories told by explorers and sailors, only the distances they have traveled, by land or sea, would be relevant; all other details would be attributed to the "creativity" of the travelers. A sufficient agreement as to the distances traveled would indicate an interest for the continents in question, but in no way prove their existence. In other words, the analysis, from a nominalist point of view, of mathematical sociologism, does not permit one to conclude anything concerning the eventual reality of the presumed explorations, but it does not rule out this possibility either.

This theory receives a rather average grade when judged by the criteria listed at the outset. The aesthetic condition is satisfactorily fulfilled, since the mathematical community is very sensitive to this component, and the same applies to the symbolic aspect. The fecundity of mathematics remains, on the contrary, a total mystery, shrouded in the psychologism surrounding creativity. Finally, the problem of the correspondence between mathematics and physical reality is left unsolved.

MATHEMATICS AND PHYSICAL REALITY

Among the criteria for comparing the various philosophies of mathematics, we have included the crucial role played by this science in the formulation of the laws of physics. It is a rather unusual criterion, often considered a minor one. We shall therefore take a closer look at it, to better put it into perspective.

There exist some "ultrafinitists," mostly physicists or computer scientists, who refuse to see in mathematics anything but a finite process executed on computers, or directly in nature by the objects themselves. It is an extreme position that we shall not elaborate on.

It is certainly true that everybody appeals to physical reality when it comes to choosing among all possible mathematical notions—not always with much conviction, though. Indeed, it is in this reality that the formalists find the source of the regularities presumably giving rise to logic, as explained by Hume. The intuitionists totally agree with them. The nominalists, for their part, tone down the excesses of their formal game by noting from time to time the suggestive influence of external reality, if only to justify the origins of Greek geometry. Finally, the logicists see in the existence of physical objects that can be made into sets the inspiration of set theory and of the natural numbers, the two pillars supporting logic.

And so a modern account of mathematics would tend to start off with logic and the existence of objects, while throwing away all other aspects of reality. Many mathematicians believe such a point of view to be legitimate, if not unassailable, and therefore that mathematics is rooted in reality's simplest elements, to evolve subsequently in an autonomous fashion.

Unfortunately, it appears that they are deluding themselves by believing in such a simple world. Perhaps one century ago reality still seemed made up of objects that one could neatly tell apart and count, but today things are no longer so simple. We now know that physical reality is governed by quantum laws, a fact that calls into question the fundamental simplicity of the very notion of an object.

One might be tempted, on the contrary, to see some very simple objects as the building blocks of physics, that is, elementary particles such as electrons, protons, quarks, and so forth. However, by a strange malediction, those objects lack all the characteristics needed to build a theory of sets based on them. These particles are absolutely indiscernible, and nothing permits us to distinguish one electron from another, not even its position in space. It is therefore impossible to say that one of them possesses the property defining a certain subset but the other does not. Present-day physics is based on objects that cannot be conceived as elements of a set from which subsets can be formed.

So what?—one might ask. I can see that on a human scale there are objects exempt from that kind of flaw: trees, stones, matches, and one thousand other things I can refer to when working out the *a, b, c* of mathematics. This is an old and respectable point of view, dictated by empiricism. There is, however, a problem with it, subtler than the previous one but no less real. If, in trying to understand the nature of these objects in a world subject to the laws of quantum physics, we attempt to provide a satisfactory description of them, we are confronted with an unexpected reversal in the supposed order of mathematical complexity. The laws of physics that translate the existence of objects can only be expressed using the most refined methods of analysis, immeasurably removed from the *a, b, c* of set theory.[5] In other words, what seemed to be a suitable starting point for mathematics now appears, in the present state of knowledge, as a particularly remote terminal point from the perspective of the theory of matter. There is nowhere in physical reality an anchorage for mathematics that imposes itself as initial evidence.

Should we let pessimism overcome us, and conclude that our persistent questioning results only in the annihilation of knowledge? We do not believe so. But this is an indication that we must be even more exacting and, above all, not consider mathematics as an independent domain but as an integral part of an all-encompassing philosophy of knowledge.

---

[5] This observation anticipates the discussion that will take place in chapters 10 and 12, concerning the classical properties of macroscopic objects derived from quantum laws.

Formal Physics

THE CENTURY OF FORMAL PHYSICS

WITH THE WORKS of Maxwell at the end of the nineteenth century, classical physics achieved a profound mutation of its nature. The leading role of the old, visual concepts—position, velocity, force—was coming to an end, mathematics having so far provided them with added precision without altering their original intuitive meaning. That clear vision had now been replaced in part by incomparably more abstract notions, that of an electric or magnetic field, for instance, whose mathematical expression was no longer a simple translation of intuition but the only possible form that was truly explicit. As a consequence, the laws of this new physics became mathematical relationships among these quantities; some laws describing their connections and others expressing their dynamics, that is, the way they evolve in the course of time. If the physicist's mind still tried to salvage as much as possible of the intuitive representation of concepts, a new era was dawning, one in which the mathematical form of physical notions and laws would take precedence over all other forms of understanding.

From now on, all of physics will rest on even more formal principles which often preclude any intuitive interpretation, when they do not openly defy common sense or what we believe common sense to be. It would be a mistake, however, to see in this amazing and in some ways frightening evolution the result of a conspiracy led by a few abstruse minds or unbridled metaphysicians, who preferred their extravagant dreams to the natural simplicity of things. In fact, no effort was spared to instill more flesh, blood, and life into notions that appeared to disallow the testimony of our senses; if they rule today, impassive, at the heart of our knowledge, it is because no one was able to dethrone them.

It would be equally wrong to believe that physics, in its race toward abstraction, has cut itself off from reality to don even thicker mathematical garments. In truth, physics reaped immense benefits, its domain of knowledge spreading out almost limitlessly until no mystery remained unsolved, no stone unturned, except at the most remote confines. At the same time, the detailed understanding of concrete reality deepened, and the technical applications proliferated. Intuition was not evicted either and still plays an active role in inspiring many ideas in physics. It only deserted the foundations, but it is enough.

The major episodes of this adventure are known, and they are few. The adventure begins bordering on the fantastic with the special theory of relativity discovered by Einstein in 1905: space and time lose the absolute character they had always enjoyed in everyone's mind, and which Newton had explicitly postulated. Distance and the passing of time depend on the motion of the observer measuring them. Shortly afterward came Einstein's relativistic theory of gravitation, providing an answer to the great question that Newton had left unresolved: the force of gravity does not act instantaneously at a distance but its effect propagates gradually at the speed of light. This major triumph is also a source of immense puzzlement: not only do space and time become intimately connected as a result of motion but together they form a new entity, space-time, totally inaccessible to intuition and, moreover, having a curvature. Only mathematics can offer a description of this new object. When confronted with it, common sense seems helpless, if not rather foolish.

We could have contented ourselves with accepting the fact that, notwithstanding the fascination they exert, space and time have never been clearly understood, and therefore any reflection concerning them is a kind of metaphysics. Also, since the new effects of gravitation predicted by the theory are minimal, we could have simply acknowledged the existence of a mysterious zone at the outer limits of physics while retaining a clear vision of matter—the matter we can still see and touch. But in so doing we would have been closing our eyes to the fact that the worst, as well as the best, was yet to come.

The best and the worst is the marvelous and at times diabolic quantum physics. But better not spoil the freshness of the subject too soon, because we shall have the opportunity to treat it at length. We can, nevertheless, reveal right away the extent of its domain. Here is its first claim: all types of matter and every form of light or radiation are composed of minute particles—electrons, protons, neutrons, photons, and a few others. Quantum mechanics is the expression of the laws of physics proper to those particles.

It is therefore the theory of everything, except perhaps space and time, the quintessence of physics, a universal theory from which the rest of physics can be derived, at least in principle. Thus, quantum mechanics is practically all of physics condensed into a few laws, and for this reason it may be said to be marvelous. We shall soon find out that it can also be diabolic, but better not to vilify it before having been properly introduced to it.

That being said, it will be impossible to discuss these theories in detail, and we shall continue to stick to the essentials, to the bare minimum for whoever wishes only to reflect about science. The special theory of relativity and the relativistic theory of gravitation will be touched upon only briefly. This is not to be interpreted as an unfavorable evaluation of their merit or intrinsic interest, but quantum physics offers enough food for thought to gratify our intellectual appetite.

As we have done before, we shall continue to examine these questions by following the course of history. Unfortunately, history is too rich and complex to be tracked in all its aspects; it is full of adventures, enlightenments, premonitions, and reversals, of big surprises too. For this reason we shall be constrained to simplify it. After establishing the boundaries of these astonishing sciences, we shall strive to extract some fundamental principles despite their mathematical complexity, in order to better grasp all their consequences. We shall then see emerge in full view some far-reaching philosophical problems where the theory of knowledge nowadays plants its roots.

RELATIVITY

Despite its formal aspects, the electrodynamics of the end of the nineteenth century had retained from the past two ideas, one strong and the other simple, which together led to an obsession. The strong idea was the absolute character of space and time. The simple one was that the oscillatory nature of light presupposes a material support: something that vibrates, known as the ether. The obsession was to prove the existence of the ether by means of some experiment.

The memory of the lucubrations, theories, and false starts provoked by this ethereal quest is today partly lost. It is nevertheless a considerable chapter in the history of science, well known only to specialists in the subject. Here are the broad outlines: It was natural to assume that the ether, a sort of material medium that vibrates at the passage of light, should be present everywhere light propagates, and in particular in the interstellar vacuum through which the light coming from the stars travels. It was therefore the materialization of the absolute space postulated by Newton.

Such an idea was not mere speculation, but based on some sensible remarks that it is appropriate to recall. There was first the composition of velocities. The existence of an absolute space and time entails that the velocity of an object, when the observer is also in motion with respect to absolute space, may be calculated with a simple rule. And so, if a light signal propagates with a certain velocity in absolute space, an observer in motion should detect the (vectorial) difference between the absolute velocity of light and his or her own velocity. The observed velocity of light should therefore be affected by the motion of the laboratory where the measurement takes place, since the earth is carrying the laboratory with it in its rotation around the sun. Now, Maxwell's equations predict a perfectly defined speed of light, designated by c, with no other alternative. It was then reasonable to think that Maxwell's equations represented the laws of physics as they hold in a very special medium, the ether, which was supposed to be at rest with respect to absolute space.

What we have just said regarding the composition of velocities shows that this conjecture could be tested, in principle, by measuring with sufficient accuracy the speed of light. Since the velocity of the earth with respect to the sun has opposite directions every six months, one ought to be able to observe a difference. It is not even necessary to wait six months, for the velocity of the earth at any given instant has a well-defined direction with respect to absolute space, and therefore the observed value of the speed of light should vary depending on whether light travels in this particular direction or in some other—a direction perpendicular to it, say. Unfortunately, there is a problem: the ratio V/c between the speed of the earth and that of light is very small, about one to ten thousand, and the measuring devices then available did not afford such precision.

Thus, other experimentally detectable effects had to be sought. Interference was a promising candidate, for the exact position of interference fringes offers the possibility of greatly amplifing the supposed variations in the speed of light. To be sure, the anticipated effect would be of second order with respect to the very small quantity V/c, that is to say, proportional to its square, and thus of an order of magnitude of one one-hundredth of a million. But the expected amplification would render the effect perceptible, provided a sufficiently large and stable interferometer was employed. It is therefore thanks to the technical advances he had made in the design and operation of interferometers that the American physicist Albert Michelson, in collaboration with Edward Morley, can finally measure the "ether wind" in 1887.

The outcome was completely different from what they had expected. All the evidence and verifications pointed to the same conclusion: the velocity of the laboratory has absolutely no effect on the measured speed of light.

Many received ideas were invalidated by this discovery, and an explanation was sought by all available means. In 1893, the Irishman George Fitzgerald came up with one that was both appealing and intriguing: Could it be that the size of material bodies, such as, for instance, the steel rod then used to represent the standard meter, would change under the effect of motion with respect to the ether? The length of a meter would then decrease when traveling in the direction of its length and remain unchanged if the direction of motion is perpendicular to it. Fitzgerald even proposed an explicit expression for such modifications. Hendrick Lorentz, a Dutch physicist and the author of a detailed theory of electrodynamics in matter, attributed this effect to an alteration of the interatomic forces in the material objects used as standards of length (as well as in all other material bodies). He showed in 1903 that the said effect should be accompanied by a change in the periodic motion inside the atoms and, on a larger scale, in the movement of clocks. Both these effects were combined in the famous Lorentz transformation, which gives the changes in length and time when two observers moving with respect to each other compare their respective measurements.

In 1905, Albert Einstein proposed a sweeping conceptual revision. Instead of assuming that motion with respect to the ether is the cause of Fitzgerald's contractions, he sees the origin of these contractions in the very nature of space and time. The same standard meter has exactly the same length for the observer holding it, and the hands of the clock he carries in his waistcoat pocket tick away always at the same pace (in 1905, observers were supposed to be men, dressed in three-piece suits). This observer measures space with his meter, and time with his clock. A second observer, traveling with respect to the first with constant speed may do likewise, but the measurements of one and the same event by the two observers need not coincide. In other words, there is neither absolute space nor absolute time; only measures of distance and time depending on the motion of the observer. The way measures taken by two different observers are related only involves the velocity of each one with respect to the other, that is, their *relative* motion.

The conceptual revolution that follows can be better appreciated when compared with Kant's synthetic a priori judgments regarding space and time. We can continue to assume, if we so wish, that these are innate ideas shaping our representation, our visualization, of the world. But it must be conceded that such categories of thought do not agree with nature, except (and this is by far the most frequent case) when all motions in question are slow compared to the speed of light. Space and time are always representable for our mental vision, but only approximately, and, ultimately, the only reliable description is the mathematical formulation of the correspondence between the observations. It is not describable by any other means than those of algebra, even if we may eventually overcome our initial puzzlement. With the advent of relativity, the theory of knowledge has forever ceased to be cast in an intuitive representation, to be based solely on concepts whose only authentic formulation involves a mathematical formalism.

But were Einstein's ideas of space and time sufficiently convincing to elicit their general acceptance? They appeared radical compared to those proposed by Lorentz, which were certainly much more classical. And so it was not from the above general considerations that the confirmation would come, but from their application to dynamics. Newton's principles of dynamics were clearly incompatible with the new theory. In order to reconcile them, it was necessary to review the way momentum and kinetic energy are expressed as functions of the velocity of the moving bodies. Thus, Newton's equations had to be rewritten, and in such a way that the new and the old formulas would significantly disagree only when the velocities in question approached that of light. This is precisely what Einstein did with extraordinary success, as is well known: there is an energy due only to mass, given by the famous formula $E = mc^2$.

The existence of this energy due to mass would later manifest itself in nuclear physics, where the binding energy between protons and neutrons in the nucleus results in a perfectly measurable decrease of the total mass. This is, of course, not the only experimental confirmation of the theory of relativity, many more being known. However, we shall not attempt to list them here, since they belong in the specialized literature. Let us rather take a look at the sequel.

THE RELATIVISTIC THEORY OF GRAVITATION

Its success notwithstanding, the theory of relativity left two significant problems still unsolved: those of knowing how to apply it when the motion of one of the observers is not uniform, and of fitting Newton's theory of gravity within the new framework.

In Einstein's first theory, a whiff of absolute space and time still remained. The Lorentz transformations that it employs only apply to two observers moving with constant velocity, that is, without relative acceleration, with respect to each other. Now, it turns out that Newton's principles of dynamics, in their simplest form, are not necessarily restricted to space and time. In fact, they keep the same form in any reference system (or laboratory within which the measurements take place) moving with constant velocity with respect to absolute space. These particular reference systems are called Galilean because they are the ones in which Galileo's principle of inertia applies, that is to say, a body not subject to any force will move in a straight line with constant velocity.

The special theory of relativity still retained the notion of a Galilean reference system. It is only in such systems that the new formulation of dynamics had the simple form assumed by Einstein. And so, even if there was neither absolute space nor absolute time any more, a certain category of reference systems was singled out as having a distinct property. Paraphrasing George Orwell, one might say that, among all possible reference systems or ways of describing space and time, some were more equal than others. This was a matter calling for further reflection.

The other problem concerned gravity. Newton himself had been bewildered by the existence of gravitational forces acting at a distance.[1] The difficulty became even more serious in the new relativistic theory, which did not allow for any physical effect to propagate faster than the velocity of light. Here is why: Suppose that there is a cause A to some gravitational effect—the ejection of a certain mass of matter by the sun, say—that is detected by an observer. The cause A produces an effect B at some distance, a tidal wave on earth, say. Can the observer affirm that the effect takes place at the same time as the cause? Certainly not, for one can show that, as a consequence of the Lorentz transformations, other observers in motion with respect to the first would see *the effect precede the cause*. The difficulties of interaction at a distance are no longer a mere metaphysical irritant, as was the case in Newton's time, but a source of internal contradictions. For this very reason, the problem they create becomes open to analysis, and this is precisely what Einstein will set out to do between 1911 and 1916.

He must face two seemingly very different questions: to reformulate the laws of dynamics in an arbitrary system of reference and, in particular, in one with nonzero acceleration, and to find a theory of gravitation where forces cannot act instantaneously at a distance.

These two problems are related, as he very soon realizes. We know that acceleration manifests itself through inertial forces: it is the force without any apparent agent but very real that we experience in our stomach while riding a roller coaster, or the one a pilot feels during a strong acceleration. We can also perceive its effect in a fast elevator. Now, while I am inside an elevator, Einstein observes, it is absolutely impossible for me to tell whether the feeling of weightiness that I experience is due to the acceleration of the elevator, to a true gravitational force, or to a combination of both. No measurement performed inside the vehicle, and without

[1] Actually, Newton and his contemporaries were puzzled above all by the fact that gravity could propagate in vacuum, the question of its instantaneous trasmission being raised only later.

131

looking outside, can settle the question. The reason can be found in a strange agreement already remarked by Newton: the equality of inertial mass (which determines the reaction to acceleration) and gravitational mass (which determines the force of attraction created by other masses).

Hence, by reformulating the laws of dynamics in an arbitrary reference system we can expect to understand what form the laws of gravity should take in a relativistic setting. Einstein will now make a crucial observation, by following in the opposite direction the path leading from the relativistic properties of space and time to dynamics. This time, he will clarify the laws of dynamics in an accelerating reference system by studying the geometry of space as seen in such a system.

The example he gives is sufficiently simple to be presented here. Suppose we are riding a merry-go-round that is rotating fast enough for the effects of relativity to be felt. We then experience an acceleration (except if we are placed right at the axis of rotation) whose inertial force is simply the centrifugal force. But is this all that is going on? With the help of a ruler, let us measure the radius of the circular platform of the merry-go-round. Since the velocity is everywhere perpendicular to the radius, our ruler is not subjected to the Fitzgerald contraction, and we thus find a certain value for the radius. Let us now measure the circumference of the platform by placing many identical (small) rulers end to end along the circumference. This time the velocity is everywhere parallel to the rulers, which will therefore experience a relativistic contraction. In comparing the length of the circumference with that of the radius we shall not find 2π, as would be expected, but a number that will depend on the radius of the platform and on the rotational speed of points on the circumference. There is ample reason for astonishment: our space is no longer Euclidean!

Thus, the geometry of space in an accelerating system of reference no longer appears to be Euclidean. Similar considerations demonstrate that the passing of time as measured by clocks is also affected by acceleration, and it is not the same for two identical clocks placed one at the center and the other at the periphery of the rotating platform.

But what does a non-Euclidean space look like? The answer is easy when dealing with a two-dimensional space. Compare, for instance, a plane with a sphere, or with the surface of a potato. The plane is Euclidean: the shortest path between two points is a straight line and the sum of the angles of a triangle is π (when measured in radians). Two-dimensional creatures living on the potato's surface might agree to call "straight line" the shortest path from one point to another, what mathematicians call a geodesic (the name reminding us of the shortest path between two points on the earth's surface). Our hypothetical creatures would soon realize that they do not live in a Euclidean space, for the sum of the angles of a triangle on the surface of a sphere or a potato is not equal to π. The space has *curvature*, as is clear in the case of the sphere.

Granted, you might say, but the potato is situated in ordinary three-dimensional space, which is Euclidean. Indeed it is, but what if we discover, as in the case of the relativistic merry-go-round, that our three-dimensional space is not Euclidean? Are we going to assume that it is embedded in an imaginary space having four, five, or more dimensions? Must our respect and admiration for Euclid stretch that far?

It is much simpler to follow Einstein and stop at the space that an observer can perceive and measure: three-dimensional, to be sure, but non-Euclidean. Beginning with Gauss, mathematicians have taught us how to describe it without having to suppose it embedded in a higher-dimensional Euclidean space, and it suffices to apply their techniques. Thus, once again, concepts that only mathematics can express, and which defy intuition, must be brought in.

One can already see the route to follow: First, the considerations concerning space and those regarding time must be combined in a single formal object having four dimensions: *space-time*.* This object will be considered as an abstract geometric space showing curvature. As a result, it is no longer necessary to resort to Galilean reference systems in order to state the laws of physics. They can now be written in an arbitrary reference system, thus breaking free from the last fragments of the Newtonian shell. The principle of inertia no longer favors Galilean reference systems, and it now encompasses the effect of gravity: a body subject *to the gravitational force only* describes a geodesic in space-time.

It remains to find a substitute for Newton's force of gravity. Since the new principle of inertia includes the effect of gravity, we no longer need to speak of force, for it is enough to be able to locate the geodesics of space-time. This can be achieved if we know its geometry, which in turn amounts to determining its curvature.

The entire relativistic theory of gravitation is therefore reduced to finding out how the curvature of space-time is determined by the mass (or rather, the energy) it contains. But according to which equations?

This final obstacle would occupy Einstein's efforts for quite some time, for the necessary mathematical methods were still unknown to the physicists. He succeeds in the end, at the same time as Hilbert, whom he had persuaded to take an interest in his research. Thus saw the light the famous Einstein's equations, which it would not be appropriate to describe here. In this respect, there is an anecdote regarding Einstein's image. For the general public, he was *the* mathematician, whereas he was above all a physicist respectful of principles and not inclined to complex calculations, which he knew well how to avoid. Didn't Hilbert say that "Herr Einstein wants to replace physics with geometry, but everybody in the streets of Göttingen knows more geometry than him"? Obviously, "everybody" referred to Hilbert's students, who were actually quite numerous. This rather acrid little story will be our excuse for not developing the formalism of the relativistic theory of gravitation, for who could pretend to put into words what proved to be a painful experience for Einstein himself? We shall also refrain from mentioning the fascinating applications of this theory to black holes or to a science finally recognized as such: cosmology, that is, the theory of space-time in its entirety or, if one prefers, the history of the universe. We had better save our energy for the last formal theory of physics, quantum mechanics, whose domain is even more extensive and its consequences more important for the theory of knowledge.

THE PREHISTORY OF THE ATOM

We now turn our attention to the microscopic components of matter and radiation, and first of all to the atoms. Atoms are present in nature from the very beginning, or at least since one million years after the birth of the universe. Let us skip those billions of years during which they were getting ready to become our flesh and bones and begin right away, and once again, with the Greek enlightenment. Leucippus, who lived a generation earlier than Socrates and of whom we know next to nothing, conceived the strange idea of the atoms—by what stunning flight of the mind we could not say. First Democritus, his disciple, later Epicurus, and finally Lucretius (a brilliant popularizer in hexameter verses) draw consequences that posterity would never forget. But let us cut this short and take a quick glance at Descartes' atoms: they are clawlike, to enable them to hook onto one another. Nothing really new so far, but here is something more interesting: in the eighteenth century, Daniel Bernoulli shows that if gases are composed of atoms, then pressure is due to their collisions against the walls of the container, which would explain why pressure depends on temperature. For this to happen, atoms must be animated by some sort of perpetual motion, similar to the one observed by the botanist Robert Brown in a drop of water seen through the microscope, where pollen grains are in constant agitation. But let us carry on.

During the nineteenth century the idea of the atom will slowly take form, thanks chiefly to the work of the chemists. In the preceding century, the distinction between simple and composite bodies had already been established. Then, Dalton and Gay-Lussac discover that chemical reactions always involve masses of simple bodies (or volumes of gas) that are in a fixed ratio of natural numbers. This could be explained by assuming simple bodies to be composed of a single type of atom, and composite ones to be made up of molecules formed of several atoms. And so, slowly but surely, the idea would gain ground, occasionally helped by a sudden breakthrough or held back by tenacious resistance. The supporters of the atomic theory were reassured by the discovery of the laws of electrolysis and, later, by the explanation of numerous phenomena in organic chemistry. The spatial configuration of molecules was beginning to be understood, but some difficult problems remained. How to explain the fact that atoms could repel each other, as happens when trying to compress matter, and at the same time bind together to form molecules? Other strange properties, known as resonance, served to fuel the criticism of the skeptics.

Toward the end of the century, new complications appeared with the discovery that atoms are not matter's extreme limit of smallness. In 1897, J. J. Thomson discovered some very light and negatively charged particles that he named "electrons." They are emitted by the cathode of an X-ray tube, and other, much heavier particles called ions come out at the other end, the anode. Could it be that atoms (whose name, let us recall, means "that which cannot be divided") were in fact composed of smaller particles, and in particular of electrons? It was an appealing idea, for Lorentz showed that it provided a clear explanation of many electrical and magnetic properties of matter. Unfortunately, once again, a major obstacle stoodout, because nothing seemed to distinguish a conductor from an insulator.

Let us mention in passing, and for the sake of the beauty of it, that in 1899 Rayleigh explained the blue of the sky as the diffusion of solar light by the molecules in the atmosphere. The rest is also worth telling: Due to this diffusion, the light from the sun loses part of its energy, particularly in the blue zone of the spectrum, and it is scattered in every direction. As the sun's rays penetrate deeper into the atmosphere, the green and then the yellow also disappear. And when the thickness of the atmospheric shield is the greatest, as is the case at dawn and at sunset, there is only a blazing of red and orange left. But the story does not end there. One day, while he was at Darjeeling, a village in the Himalayan foothills where the English sahibs liked to take refuge from the rigors of summer, Lord Rayleigh noticed that the frozen slopes of Everest, a few hundred kilometers away, appeared in a greenish hue. From the thickness of the atmospheric layer the light should have penetrated to reach the mountain, he deduced Avogadro's number, that is, the number of atoms in a given mass—for instance, the number of hydrogen atoms in 1 gram of hydrogen, or of oxygen atoms in 8 grams of this gas. This number is twenty-four digits long, which shows how small atoms are. In order to measure their size a special unit of length is used, the angström, equivalent to a ten-billionth of a meter.

In 1905, Einstein—still he—took up the question of Brownian motion, the erratic movements of pollen grains in a drop of water that we mentioned earlier. This motion, he says, arises from the multiple collisions of water molecules against the grains or any other tiny particle. Pursuing his idea up to its quantitative consequences, he was able to predict the average distance traveled by a particle in a given interval of time. His prediction was soon to be experimentally confirmed by Jean Perrin, and it is customary to consider this event as marking the general recognition of the existence of the atom.

Much more would be learned on the nature of atoms in 1911, thanks to Ernest Rutherford. He was working on experiments where alpha particles produced by the disintegration of radium passed through a thin metal sheet. The particles were observed to depart slightly from their initial path. Since alpha particles are charged, the effect might have been due to electric forces, but the electrons in the atoms were too light to account for the deviations. Rutherford then analyzed the data and showed that the only explanation was the presence, at the center of each atom, of a positively charged "nucleus" where almost all of the atom's mass was concentrated. This was the first satisfactory model of the atom: a nucleus surrounded by electrons. Now, the nucleus exerts an electrical force of attraction on the electrons, similar in form if not in magnitude to gravitation. In the resulting portrait of the atom, vacuum taking up by far the largest part, with electrons revolving around the nucleus, lies the perfect setup to apply the well-known methods of mechanics.

How beautiful and how simple! one might have exclaimed. But hard reality would soon catch up. The history of the atom was long marked by the intrusion of new problems each time progress appeared in sight. Rutherford's model did not provide any explanation for the chemical properties of molecules, but the worst problem was elsewhere. An electron gravitating around a nucleus must accelerate, and it had been known since Hertz' time that an accelerating charged particle emits electromagnetic waves. Such should therefore be the case for the atomic electron, and with a particularly strong effect, for the electron is subjected to a tremendous acceleration inside the atom, its mass being very small and the electric forces involved considerable. A simple calculation led straight to disaster: in a fraction of a second the electron should radiate desperately fast, and at the same time travel at full speed toward the nucleus to compensate for its loss of energy. Thus, the atom ought to collapse almost instantaneously. There was obviously an error somewhere, but no one could find it. Unless . . . unless the usual laws of physics themselves collapse and are no longer valid on the atomic scale.

It was not an entirely implausible hypothesis, for something similar had happened a decade earlier regarding a completely different problem: the radiation spectrum of a black body. What physicists call a black body is simply any black object, with no proper optical emission. It can be observed that such an object emits a radiation whose color, that is to say, its spectrum, depends only on the temperature of the object. This effect manifests itself when a metal or a piece of coal becomes red at temperatures in the hundreds of degrees, and then appears brilliantly white when the temperature reaches several thousand degrees. What is called the spectrum of radiation is the distribution of luminous energy emitted as a function of the frequency (or the wavelength). In this case, too, physicists had believed they could solve the problem theoretically and compute the spectrum using the laws of thermodynamics. But such a line of reasoning led to an absurdity: a chunk of coal should emit light of infinite intensity!

In 1900, Max Planck had found what we may irreverently call a "trick" or a "ruse" to circumvent the difficulty. He had assumed that coal atoms do not radiate continuously, as would be expected according to electrodynamics, but that they emit "puffs" or quanta[2] of radiation, the energy emitted in each puff being proportional to the frequency. More precisely, he had supposed that the energy so emitted was equal to the product of the frequency with a number now called Planck's constant, a very small number on a human scale. It is a remarkable fact that so simple a hypothesis, even if incomprehensible at first sight, leads to a perfect agreement with everything we can observe and measure.

Since Plank's hypothesis had resolved a difficulty concerning radiation, it was not unrealistic to expect that a similar idea might explain the absence of uncontrollable radiation in Rutherford's atom. It remained to find a way to bring Plank's constant into the picture.

CLASSICAL PHYSICS IN A STRAITJACKET

The honor of finding the solution fell on Niels Bohr, a young Dane then working under Rutherford. By bringing together Plank's and Rutherford's results, Bohr proposed in 1913 a new dynamics of the atom destined to fame and glory. It was a conservative model, in the sense that it modified classical physics as little as possible—in fact, the model preserves all classical laws, imposing only one additional condition. Bohr studied in particular the hydrogen

[2] The word "quantum," of Latin origin, means simply "a certain quantity measured by a whole number."

atom on account of its simplicity. This atom possesses a single electron, which, according to Kepler's laws, must "gravitate" around the nucleus on an elliptical orbit. To prevent the collapse of the electron, Bohr assumed that only certain orbits can actually be described, and that the electron cannot radiate while traveling along the smallest of these ellipses. When the electron does radiate, it emits a puff, a quantum of luminous energy.

How are the permissible ellipses to be selected? The answer is in fact quite simple, because it only requires introducing a condition without new data except Planck's constant. Let us say, without writing any formulas, that there is essentially only one possible rule and it is the one postulated by Bohr. In short, he adds one more law to the purely classical theory of the atom, a law in which Planck's constant occurs. Bohr's solution turns out to have spectacular, far-reaching consequences: each permissible elliptical trajectory must correspond to a well-defined energy, whose expression involves the electron's mass, its charge, and Planck's constant, together with an integer, called the quantum number, that labels the successive ellipses beginning with the smallest one. Bohr also postulates that in order to radiate, the electron must suddenly move from one ellipse to another of lower energy level (these are the famous quantum jumps) and that the energy liberated in this way is related to the frequency of the emitted light by Plank's relation. Using his theory, Bohr then predicts the radiation frequencies that the hydrogen atom can spontaneously emit. These frequencies make up what is known as the spectrum of atomic radiation, already observed and measured long before Bohr's time. The agreement between model and measurements is very good.

Many were enchanted by this beautiful result, and foremost Einstein. The general direction it pointed at could only please him: it retained the advantages of known physics enriched by a new acquisition, a condition capable of selecting the possible states of the atom that preserved the essential features of the old model. In short, physicists increased the list of known laws, but without really modifying their way of looking at things. It is this idea of a familiar physics subjected to new constraints that we may call, somewhat cavalierly, classical physics in a straitjacket.

A period of intense research followed, dominated by the contributions of Bohr and Arnold Sommerfeld. Efforts were directed at extending hydrogen's success story to more complex atoms.

Unfortunately, the results were disappointing, and, as the theory continued to be refined and the experiments multiplied, there were plainly more defeats than victories. For example, the effect of a magnetic field on the frequencies of an atomic spectrum (Zeeman effect) or that of an electric field (Stark effect) led to the worst possible situations. The results were satisfactory for certain spectral lines of some atoms and totally false for others, for reasons no one could explain. As for the links with chemistry everyone had been waiting for, they remained as elusive as ever. If the classification of atoms in Mendeleyev's table was partially understood, there was still no sign of how chemical properties were determined. In a nutshell, physicists stalled.

The Assassination of Classical Physics

Shortly before these events took place, the French poet Arthur Rimbaud had seen "the time of the assassins" coming. Here they are now closing in, those youths, some of them very young, who will not hesitate to trample down their ancestors' legacy. Too bad for Newton and Maxwell, too bad, if need be, for the common sense gathered through the centuries by the entire human race. The time had come for physics to rise to the occasion, to come up with a theory, whether clear or not, simple or otherwise, it does not matter; but one that should account for the facts, all of them.

Those well-mannered young people certainly did not have the anarchic intentions we seem to lend them. They were searching for an honest solution in all sincerity, and it is certainly not their fault if this solution would turn out to be such a revolutionary one, as the future would prove.[3] But let us begin with the introductions: This is Louis de Broglie, member of an ancient family, trained in history, ill prepared for the role he was to play, and who contracted a passion for physics through his brother Maurice, an X-ray specialist. He is thirty-one when he begins working in physics. Over here, Werner Heisenberg, with a passion for physics, trained in German schools and particularly fond of classical Greek. At the

[3] It has been pointed out that this period following the First World War also witnessed other calls into question, such as the one brought about by surrealism. But this fact does not affect our main argument.

time of his first major contributions he is only twenty-two. Next to him his Austrian friend Wolfgang Pauli, a precocious genius who has written a remarkable survey article on the theory of relativity before the age of twenty. Here is also Paul Dirac, an Englishman as young as the others and coming from the prestigious Cambridge University. And over here the senior ones, some ten years older than the young generation: Max Born, a German universal physicist, formerly one of Hilbert's assistants at Göttingen at the time the master was interested in physics; Erwin Schrödinger, another Austrian, a student of Sommerfeld who is still uncertain of his true calling, but whose sound mathematical training will come in handy. Watch them as they enter the arena under Einstein's and Bohr's attentive look, ready to encourage or correct them, as the case might be.

Heisenberg opens the discussion, and we shall listen to him for a moment. He does not hesitate to call into question the foundations of classical physics, challenging most of its concepts. Are we certain, he wonders, that the notions of position and velocity also apply to objects such as the electron? It is impossible to know exactly where the electron is inside the atom, for we would have to use some kind of device that could only result in the destruction of the object of our observation. Could it be that, besides this practical impossibility of determining position, the laws of physics would rule out its very idea, and that we would be allowed to employ only concepts that can be experimentally verified? In asking this last question, he believes he has been inspired by Einstein, who, in his questioning of the traditional conceptions of space and time, accepted as valid notions only those that can be measured in the laboratory with rulers and clocks.

But if we must do without the classical ideas of position and velocity, what shall we replace them with? asks Heisenberg. Since the support of visual intuition must be given up, he will resort to formal concepts. But first, he has to find the mathematical objects that will replace the familiar notions. Heisenberg then undertakes a reflection that would be impossible to follow without a considerable technical background. Here is a broad outline: The acceleration of an electron at the origin of radiation only manifests itself at the time of a quantum jump between two (quantum) states of the atom. Thus, the object "acceleration" is certainly not a number, as we usually envisage it, but depends on the initial and final states of the atom at the very instant the jump takes place, and it has a meaning only at that time. Therefore, if we enumerate the atom's possible states as Bohr did with his "quantum numbers" that label the energy levels, acceleration then becomes a quantity that depends on the quantum numbers of the initial and final states. Thus, acceleration may be replaced by a double-entry table of numbers, labeled by integers indicating the initial and the final states. Heisenberg is led to analogous considerations for position and velocity, which he replaces in a similar way by tables of numbers. He then succeeds in reformulating the essentials of the laws of mechanics using such tables. In 1924, Max Born, to whom Heisenberg has confided his discoveries as well as his bafflement, encourages him to publish his results, after informing him that his tables are called matrices by the mathematicians. With the help of Pascual Jordan, an invaluable recruit given his knowledge of these yet-little-used mathematical entities, he soon creates an almost complete version of a new mechanics, accompanied by a great number of predictions and results, all of them as valuable as they are convincing. This new theory is then refered to as matrix mechanics.

Not long before, in 1923, Louis de Broglie had published an altogether different idea, but which would bear its first fruits only after the publication of Heisenberg's results. This is why, at the time, the two contributions were discussed in reversed chronological order. De Broglie's idea was based on earlier work by Einstein, who had interpreted Planck's quanta of luminous energy as well as the characteristics of the photoelectric effect (where electrons are emitted by a metal under the action of light) as being due to the existence of grains of light: the photons. And so Einstein reactivated the old idea of the corpuscular nature of light, and managed to demonstrate, however incompletely, that the existence of photons did not contradict interference phenomena. Light, which ordinarily manifests itself as a wave, is made up of particles. Why, asked de Broglie, couldn't this idea be turned upside down and generalized by assuming that every elementary object, an electron, for instance, which normally appears to be a particle, is associated with a wave that we have not yet perceived, with a *wave function**
that we do not imagine?

Several years would have to elapse for this idea to be experimentally confirmed by diffraction effects, that is, by purely undulatory effects, analogous to interference, produced by electrons traversing a crystal. De Broglie's idea had previously been communicated to Einstein, who discussed it with Sommerfeld. The latter then posed his disciple Schrödinger the following problem: how to compute Louis de Broglie's wave for an electron inside an atom, and how to formulate this wave's dynamics, that is to say, the way it evolves as a function of time?

Almost immediately, in that same year 1926, Schrödinger finds a solution to the problem and proposes an equation for the wave's dynamics that will be named after him. He puts it to the test by computing the hydrogen atom spectrum and the results coincide with those of Bohr. He also accounts for a number of subtler phenomena that could not be explained by either Bohr's or Sommerfeld's methods, and in particular he manages to reproduce Heisenberg's results.

Can the meaning of this magic *Schrödinger's equation** be put into words? It does not seem so, and we are once again immersed in a formal physics. It is impossible to say, as Voltaire used to do about Newton's mechanics: "You see, acceleration is the relation between force and mass, and the rest is only a matter of calculation." We can only say that Schrödinger's equation involves the electrical energy of the interaction between the nucleus and the electrons, as well as among the electrons themselves. The masses of the electron and the nucleus also occur in it, the latter playing a lesser role. It is not a very appealing equation at first sight, although it is rich in subtle properties; a partial differential equation, that is, one involving partial derivatives, vaguely resembling Maxwell's equations. What can we add? Nothing, really, except to write it down, but it is of course out of the question to do it here.

The new mechanics would immediately prove fertile in results, which were almost invariably fully consistent with experience, and never at variance with it for too long. The curse that had long afflicted the infancy of the new physics was finally exorcised, and the time when each step forward generated new difficulties seemed forgotten like a bad dream. This time, every obstacle would soon be a source of progress, carrying with it the explanation of many other phenomena. Everywhere, in concert, Schrödinger and de Broglie's wave mechanics and Heisenberg's matrix mechanics made a hit, with identical results, despite the fact that they appeared to be so different.

Indeed, Schrödinger did not use any matrices, and waves had no place in Heisenberg's system. Which of the two theories would have the last word? Strangely enough, neither of them, for still in 1926, Dirac and Schrödinger both showed that the two seemingly different mechanics are in fact one and the same theory, and that it is possible to transform one into the other by purely mathematical manipulations. It was then decided not to favor either of them over the other, and to call the synthetic theory so obtained quantum mechanics. Nowadays, the choice generally falls on a more abstract version of the theory principally due to Dirac and von Neumann. But it will not be necessary for our purposes to consider it, since its interest is mainly technical and it does not involve any essentially different principles. Yet another version, also equivalent to the others and shedding a new light on some aspects of the theory, would be proposed at the beginning of the 1950s by the American physicist Richard Feynman. But the nuances and subtleties of the various translations of the Bible of Nature are of interest only to the theologians of physics.

The Harvest of Results

Before presenting the principles of quantum theory and their impact on the philosophy of knowledge, it is perhaps appropriate to give some idea of the extent and fecundity of the theory, if only to justify the importance attached to it. We shall therefore very briefly and incompletely review its impressive harvest of results without paying much attention to dates or chronological order, for applications began to develop on several fronts at once beginning in 1927. That year the theory reached its maturity, and was to remain practically unchanged up to this day.

Let us begin with chemistry, which was after all at the origin of the whole story. Schrödinger's equation enables us to compute the wave function of the electrons in any atom or molecule (actually, the necessary calculations are extremely difficult, and only feasible after the advent of computers). The possible energy states of a molecule can therefore be calculated to determine which ones may exist when two or more atoms are bound together. The position of those atoms inside the molecule can be known too, and in many cases even the nature, the efficiency, and the speed of the chemical reactions may be computed. The whole of chemistry became thus accessible: first, of course, by the perfect understanding of Mendeleyev's table, but also by the full comprehension of the structure and the chemical properties of molecules (the helicoidal shape of the DNA molecule, for instance). The formerly puzzling phenomena of resonance and the changes in form that certain molecules underwent became clear, even though the reason for them remained buried in the formalism and not expressible by simple images. At present we can determine the expected properties of a new molecule before actually constructing it, and only verify these properties experimentally afterward. If chemistry thus experienced a new boom without losing its original character or its own specific techniques, it is also true that its foundations are now indistinguishable from those of physics.

The physics of ordinary matter, and in particular solid state physics, has also been profoundly transformed. Quantum mechanics finally provided an explanation of the difference between conductors and insulators. The same applies to the properties related to heat (heating capacity, phase transitions, heat conductivity), optical properties (transparency, color, refractive index), magnetic properties (those of the iron in an electromagnet, for instance), and mechanical properties (hardness, plasticity). The explanation of superconductivity phenomena obtained in 1958 marks the end of the heroic era. But this did not mean the end of the harvest. Every year we witness some remarkable developments, although these are mostly geared toward the systematic application and refinement of known results, to which should be added the study of more complex phenomena (liquid crystals, surface phenomena). Among the most outstanding practical achievements we find the invention of the transistor, thanks to which the computer boom became possible, and the recent discovery of superconductors at relatively high temperatures.

Optics greatly benefited from the invention of the laser, which is based on a purely quantum phenomenon: the fact that the presence of photons near an atom may stimulate the emission by the latter of new photons. From a more fundamental point of view, optics succeeded early on in reconciling the undulatory character of light with the existence of photons, even if, again, those results cannot be put into words.

The fallout from quantum mechanics would also have an impact on two totally new branches of physics that study, respectively, the atomic nuclei (nuclear physics) and elementary particles. Both have experienced a stunning development, beginning in the thirties in the first case, and from the 1950s on in the second. We shall not say any more, though, for it is not our intention to give a full report. Let us only add that, despite the many efforts to discredit it, quantum mechanics has always come out on top, and that today it may be considered as a completely accurate theory, even when experiments involve distances between particles of one-billionth of an angström, or energies thousands of times that of the proton's mass energy. The agreement between theory and experience has in certain cases reached over ten significant digits, a precision unequaled in any other scientific domain.

In fact, the above-mentioned results can give only a pale idea of the abundance of the contributions this prodigious theory has borne. Directly or indirectly, all of physics and chemistry depend on it, and, as a consequence, all that nature has to offer. It is a true Aladdin's treasure, and we shall later see what kind of genie he can conjure up with his oil lamp. One thing is certain, though: the language he speaks is a formal one, and it is not our language.

The Epistemology of Physics

As we have done for mathematics, we must examine the state of the philosophy of physics after this science turned formal. While a philosopher can cope with the technical difficulties of the theory of relativity, the obstacles presented by quantum physics are considerable, which may explain a certain convergence of the views held by physicists and philosophers on its epistemology. In the thirties, the greatest names in physics participated in the debate, and until recently the interested philosopher was reduced to discussing the opinions of Bohr, Einstein, Schrödinger, Heisenberg, Pauli, de Broglie, and a few others.

I have not tried to cover the whole field but rather focus on its most significant examples. Therefore, we shall stay away from everything concerning space and time—actually, many authors devote the bulk of their texts to a discussion of Einstein's views on the subject.

To be fair, some recent works deal with the latest results, which would be excellent news were it not for the fact that the authors often embark on wild speculations. This is a dangerous trend, as much for the public at large as for philosophers, who may have difficulty finding their way, especially since some reputable physicists are among the writers. They certainly deserve praise for reporting on the current avenues of research, but they mislead the reader by not sufficiently warning him or her that these avenues may well lead nowhere. It is so easy to dream with the help of some mathematics in the wonderland of general relativity.

I shall therefore restrict myself to quantum physics and be brief, for we shall see in part three how some recent progress of the theory shed a new light on its epistemological consequences. For the time being we shall limit ourselves to what is generally admitted, by highlighting the outstanding features of Niels Bohr's masterly theory. I would like to demonstrate that Bohr had no choice but to establish this framework at the outset of the theory, and that he was obliged to impose strict and very restrictive rules of thought.

The formidable interdictions he promulgated were beneficial to physicists, by preventing them from raising questions that might perhaps have stopped them (strict rules make the work easier, as any author of sonnets would confirm). Unfortunately, such rules also resulted in a great philosophical confusion.

We are today in a better position to decide what to retain and what to reject from Bohr's masterpiece. Not much needs to be altered from the point of view of the working physicist, but regarded from a philosopher's perspective the theory requires considerable modifications. Bohr's work remains nevertheless too familiar a reference to deserve anything less than a whole chapter, if only in order to understand why it had to be built the way it was.

WHY DO WE NEED INTERPRETATION?

Let us look at an object, any object, even the most ordinary one, a billiard ball, for instance, and let us compare the way we think of it with how contemporary physics describes it. Nothing could be simpler for us: everybody has seen such an object, and, as you were reading the first sentence of this paragraph, an image of the ball formed in your imaginaton. Less than a century ago, a physicist would not have thought otherwise, except for some additional precision—by associating the coordinates of the ball's center with numbers, perhaps. To an atomist, the ball would have appeared as a dense pack of atoms, with each atom being imagined as another, very small kind of ball.

There is nothing resembling all that in quantum physics. The physicist does start from the idea of the ball as an assemblage of a colossal number of atoms, but this idea is immediately replaced by a wave function which depends on as many variables as there are electrons and atomic nuclei in the ball. His or her notion of the center of the ball's position is not very different from that of the classical physicist. But to talk about velocity he or she first needs to differentiate the wave function with respect to certain variables, then divide it by the complex number i (the square root of -1), and perform many other complicated calculations before finally declaring, "I cannot tell what is the exact velocity of the ball (no more than the exact position of its center), but here is a probability distribution of the velocity having such and such a value." The contemporary physicist has no longer any precise image of the ball left, at best only the blurry impression of a cloud of probabilities.

And yet the ball appears to be there; it rolls. Everything seems indisputable: the atomic nature of matter and the quantum laws governing the particles, which have been confirmed by every experiment, the impossibility of reaching or conceiving by means of the theory anything but probabilities. But equally indisputable is another fact: the ball is there. Had it been capable of laughing, the ball would certainly have a good laugh at our expense. We do not understand, we no longer understand, the alpha of fact seems to contradict the omega of theory.

The purpose of *interpretation** is to reconcile these extremes; to show, if possible, that they are coherent; to establish modes of thought capable of bringing them together without deforming them. It is hard to imagine a more philosophical enterprise, for it boils down to knowing how to think about the world.

There are at least two ways to envisage interpretation. One is based on humanity's shared experience, its representation of a world filled with facts, its ancestral common sense. This approach selects those things that are compatible with the discoveries of physics, purifies the concepts, limits their scope, and finally talks about the world with the caution of a cat. This is the path followed by Bohr. Another conception would consist in viewing interpretation as a particular branch of theoretical physics. Beginning with certain given principles (the existence of particles, wave functions, etc.) one would deduce, through mathematical demonstrations, the features of the classical, common sense representation of the relatively large objects we perceive on our human scale. This is the more recent approach, which we shall discuss shortly.

It is clear that the wrong (if we may call it so) that the interpretation seeks to right stems from the formal character of science, from the fact that its initial concepts are not accessible to the imagination. Given that all of physics is more or less formal, including what we call classical physics, there is everywhere a need for interpretation. This is hardly noticeable in Newton's physics, it becomes already a slight embarrassment for some (the shrewdest) with Maxwell's electrodynamics, and it is definitely manifest in the theory of relativity. In the last domain there is, however, a simple method to obtain an interpretation: to imagine here and there, wherever necessary, observers in motion. The device is so convenient that many people do not realize that the goal of those imaginary observers is to provide an interpretation.

In quantum mechanics, interpretation is essential for at least three reasons: First of all, because the formalism of the theory could not be more obscure; secondly, because the very notion of an observer is no longer clear, and those who have employed it ended up by implicating the observer's conscience, in contradiction with the objective character of science;[1] finally, because the probabilistic aspects of the theory must eventually be reconciled with the undeniable existence of facts, and so the interpretation ceases to be a mere translation and becomes a theory in its own right.

UNCERTAINTIES

One of the most striking features of quantum mechanics is its probabilistic character. Everything in the quantum world occurs at random and there is no direct cause for quantum events. Moreover, the probabilities of such events strongly differ from the probabilities used in classical physics since Laplace's time. These can be explained roughly as follows. Everything obeys the laws of classical physics and every event has a cause, some mechanism acting somewhere. An apple falls from a tree because its stem weakens, the wind blows, or a bird hits it. We cannot tell exactly when the fall will occur, but some direct mechanical cause is in action. If we knew the state of the fibers in the stem and its exact evolution in time, we would be able to tell when and why the apple will fall. But we do not know, or do not care to know, which is why we resort to probabilities: They express a reasonable expectation despite our ignorance of hidden details. In a nutshell: Everything in classical physics is determined, and the use of probabilities is only a substitute for the exact knowledge of the acting causes.

Things are very different in quantum mechanics, for in it events really occur at random. No cause is at work to make an excited atom decay at some specific moment. There are, of course, laws governing the whole process, but they only express the probability of the event taking place at one time rather than another. Quantum

---

[1] An objective science, according to Kant, only refers to objects independent of the mind.

probabilities are not a substitute for a detailed knowledge of hidden, relevant details; there are no relevant details, just pure chance.

This idea was introduced by Max Born and has been fully confirmed by innumerable experiments (notably recent ones, where a single atom caught in a trap and subject to a laser beam is seen to continuously produce a fluorescent radiation, except when unmistakable "quantum jumps" occur). Born gave explicit rules for computing quantum probabilities in terms of the wave function, and these have always shown an excellent agreement with experimental data. Einstein was nevertheless horrified by the idea: "God does not play dice."

A conceptual abyss seemed then to separate classical from quantum physics, determinism from pure probabilism. The most puzzling feature of this apparently irreconcilable opposition is that both opposites are necessary for physics. Probabilism is an essential trait of quantum physics, and is fully confirmed by experiments. But how are experiments carried out? They involve some laboratory equipment, measuring instruments, and so forth. But regarding any of this apparatus we may ask: Why do we trust it, why is it called a laboratory instrument? The answer is that it works as expected and predicted, provided the right buttons are pushed; in one word: because it is deterministic.

One may go so far as to recognize determinism as a necessary condition for any experimental verification of quantum probabilism. As a matter of fact, probabilities predicted by quantum mechanics are checked experimentally by comparing them with relative frequencies in a large collection of data. This comparison can only be made by gathering enough data, and for this we must rely on records of all the data that are contained on some support—yesterday a notebook, today a computer memory. It should be obvious that we consider these records as reliable witnesses of past facts that occurred when each individual datum was recorded. But if these records are trustworthy it is because each entry made at that time completely determined the state of the present record.

Physics stands on two legs: theory and experimentation. Theory requires pure probabilism and experiments can only make sense if something essential in their workings is deterministic. The untying of these seemingly contradictory demands has been one of the greatest achievements of recent theories, and it will be discussed in a coming chapter.

Probabilism has yet another consequence, which underscores some of the most formal aspects of quantum theory: Heisenberg's famous uncertainty relations. When applied to a particle, these relations say roughly that the price for more precision in the particle's position is less precision in its momentum, and vice versa. They are a direct, indisputable consequence of the basic principles of the theory.

Any Greek philosopher or part of our mind with a remnant of Greek thought would repudiate such a statement. Why? Because momentum is proportional to velocity. Suppose I try to "see" a particle in my mind. It has a position, which moves with some velocity along a trajectory. But if a sharp position makes the velocity fuzzy, there is no trajectory. I can no longer see. If Aristotle was right in saying that understanding something begins with having a clear picture of it in the mind, one may well wonder what is going on. The apparent irrationality of atoms may be told with some clumsy couplet, such as: Formal sciences make blind, unreal with a fool's mind.

THE PRINCIPLE OF COMPLEMENTARITY

The principles of the theory associate physical magnitudes with certain mathematical objects, the *operators*,* one of whose main properties is not to commute with one another. Without going into the details, let us say that this fact is the formal origin of Heisenberg's *uncertainty relations*,* which prevent us from simultaneously attributing to a particle a well-defined position and a well-defined velocity. In a similar fashion, we cannot describe light as being at the same time both an electromagnetic wave and composed of photons.

Bohr's first major contribution to the interpretation originates in the above impossibilities. I can speak of an atom's position or of its velocity at a given instant, he says basically, but I must make a choice. These manners of speech, these descriptions, are *complementary*. By this I mean that each one of them is in itself correct, with no internal contradictions, but that it is impossible to combine them. From the standpoint of common sense, this is certainly very strange, as if it could be said about someone we shall call "it":

"When I talk to 'it' on the telephone, it speaks like a man, but when I see 'it,' it does not speak at all and looks exactly like a cat." "It" could be any atom, any electron, or light itself. Rather than hearing or seeing them, particles or waves are detected (through interferences). Is "it" a wave or a particle? How can it exist in each of these forms but never in both at the same time? The impossibility of combining such exclusive features was put forward as the first principle of interpretation, the *principle of complementarity.** Bohr was so convinced of the significance of this principle that he later searched for other examples of it, in philosophy as well as in biology and psychology. Surprisingly, he does not seem to have been aware of the notion of universe of discourse, which logic had elucidated long before.

The principle of complementarity carries with it two immediate risks. The first one is the threat of paralogisms: how is it possible to remain logical and coherent when the same object may be envisioned in two, or even one hundred, different ways? The second danger lies in the arbitrariness of choice: according to which criteria should I favor one description over another if not by my free will, I who think and speak, and so risk betraying objectivity? Bohr's reply is unequivocal: we must not even mention atomic objects, he says, and only use the formalism for what it offers us— numbers, probabilities. Let us not talk about those atomic objects and give this interdiction the status of an imperative rule.

Let us consider for a moment this injunction, this commandment: "Thou shalt not talk about the atomic world in itself." There would be others, but this one is typical of the direction in which Bohr was taking the interpretation. We can still keep the ordinary representation of the world, but its scope should be considerably restricted.[2] There exist forbidden things. We cannot avoid thinking of Kant and of the peculiar fate of reason, always troubled with questions which cannot be ignored, because they spring from the very nature of reason, and which cannot be answered, because they transcend its powers. Bohr, like Hume before him, and for reasons that are not unconnected, pronounces an interdiction and imposes the existence of the inaccessible, of the unthinkable. We may then consider Hume, who renounced knowledge of the origin of the

[2] This is exactly the same approach Bohr followed when he formulated his famous 1913 model.

world's order, Kant and his unsolvable antinomies, and finally Bohr, as the great princes of the forbidden.

What are we then allowed to think, positively, according to Bohr? He says it clearly: We shall talk only about the things we can see and touch, that is to say, in the circumstances, the instruments we use in physics. We shall disregard the atomic nature of the matter those instruments are made of, together with the corresponding quantum laws. We shall take into consideration only the facts, without any mental reservations. Yes, the things I see are such as I see them: they are classical, and I forbid talking about them in any other way. Those who dare challenge this proscription be warned: they risk the worst disappointments and the disintegration of thought.

Bohr also explains the reasons that led him to adopt such a position. When talking about classical physics, he does not really mean Newton's mechanics or any other scholarly product of the intellect. His roots go deeper, down to what is clearly and perfectly representable, to the only ground where he believes it possible to state a truth, to remember the past and record facts, to reason and think with certainty. His reasons belong to the domain of classical logic, the most reliable one by human standards. The classical path is chosen because it is the only one, or so it seems, to permit a logical apprehension of the world.

As an immediate benefit, Bohr can easily remove the arbitrary component implicit in the principle of complementarity: we shall only talk about the atomic quantities that are directly revealed by some measuring device. How am I going to talk of a weak electromagnetic radiation, for instance? I simply won't, unless it has been detected—recall the interdiction on talking about the quantum world in itself. If it so happens that the radiation has been detected by an antenna, then, no problem, we can then speak of a wave or, if desired, of an electric field, because this is what an antenna measures. If the radiation is detected by a photomultiplier, a photon counter, we may then speak of photons.

Bohr's solution had the merit of being expedient, since it allowed physics to pursue its path toward new discoveries. But on the other hand, it raised a formidable difficulty. For physics now appeared to be split between two opposing systems of laws: the classical one, deterministic and a haven of certainty, and the purely probabilistic quantum system, with its complementary possibilities, at the mercy of chance and uncertainty. How can one and the same man serve two masters, one and the same science obey two categories of laws? By clinging to facts and proclaiming them to be the only truth Bohr had opened a logical breach, and a particularly dangerous one, for it threatened the very coherence and unity of science.

Many refused to close their eyes to the fundamentally quantum nature of matter to retain only its classical appearance. Schrödinger's cat, which we shall meet later, is an illustration of such a position, and von Neumann's attempts to build a quantum theory of measuring devices sounded a warning to physicists, just as the famous cat alerted the nonspecialists. Einstein never could bring himself to follow Bohr, and his mistrust even led him to question the intrinsic random character of quantum phenomena. Louis de Broglie and Bohm tried to come up with other theories to modify or complement quantum mechanics. Einstein, Podolsky, and Rosen, Bohm again, and later Bell sought different ways to put the principle of complementarity to the test in particularly subtle situations. To the very end, Bohr remained impassive in the face of all those attempts.

THE REDUCTION OF THE WAVE FUNCTION

One may wonder whether there is still a need for wave functions, given that we cannot refer to them when talking about experiments. To be sure, Bohr does not disallow the quantum formalism, but he restricts it to its role of computing device, of foreteller of probabilities. That's what a wave function is: the fuel of a machine that manufactures probabilities. Quantum theory is true, not as a fact but as a set of rules that interrelate facts, rules that have been corroborated by experience in terms of relative frequencies of the quantities that are measured. The notion of relative frequency is that of ordinary probability theory, for example, the proportion of the occurrences of the number 12 in a large number of spinnings of a roulette wheel. Instead of a roulette, we have atoms; instead of the combinatorial calculations of the kind Pascal used to carry out we have others, involving the wave function.

All right, but isn't there a snag here? How can we know the wave function of an atom when we are not supposed to talk about the quantum world in itself and pretend to know only classical data?

Bohr clears this hurdle by introducing a new rule. We must consider, says he, not only the measuring instrument but also the generating device, the particle accelerator as well as the particle counter. Often, the mechanism that creates is practically indistinguishable from a measuring instrument, and this is the case that interests Bohr the most, almost to the point of making it the general rule. He considers as particularly important the situation where there are two consecutive measuring devices. If we know the wave function of the atom coming out of the first instrument, we will then be able to predict the probabilities of the possible results of the second measurement, thus allowing an experimental verification of the theory.

Bohr defines this wave function by means of a special rule, the "reduction" of the wave function. It is a technical rule that we shall not specify, but which boils down approximately to this: tell me the result of the first measurement and I shall give you the wave function with which to compute the probabilities of the outcomes of the second.

This raises a semantic question. What does this prescription mean? It might only be a kind of rule of thumb or empirical rule, giving the so-called conditional probabilities, that is, the probabilities of the various results of the second measurement under the assumption of given outcomes of the first. This information could then be presented as a double-entry table, or matrix, with as many rows as there are results for the first measurement, and as many columns as the possible results for the second. Each entry would then indicate the relative frequency of the occurrence of the two results in succession, and Bohr's rule would merely specify how to calculate these numbers. If the reduction rule is simply such a prescription, it might be deducible from the fundamental principles of the theory, and this is precisely what has recently been done. But Bohr did not think such a deduction possible, and he took a completely different approach. After the first measurement has taken place, he assumes that the atom's wave function suddenly loses all memory of what it might have been in the past, to actually become, instantly, what the rule prescribes. The rule is then a law of physics unlike any other. Without the rule, we could not know the wave function or compute the probabilities, and the comparison of theory and experience would become impossible. The reduction of the wave function is then a sine qua non for experimental semantics.

The deep cleavage that Bohr had already dug into the heart of physics with his two categories of laws, the classical and the quantum ones, has now become even wider. If we try to imagine the measuring instrument as composed of quantum atoms and admit that the same Schrödinger's equation describes both the atom and the device that measures it, we find that the reduction proposed by Bohr is mathematically incompatible with the famous equation. Thus, the reduction forces us to do more than simply close our eyes to the atomic nature of the instruments: we must reject it. Strange situation, especially since countless experiments, some extremely precise and others very subtle, all agree on one point: the wave function reduction rule, at least as represented by a double-entry table, is thoroughly verified.

The dilemma is therefore this: Are there in effect two categories of physical laws together with a very strange reduction phenomenon not belonging to either of the two categories? Or is there only one category of laws, by necessity the most general ones (that is, the quantum laws), and the reduction rule is simply a direct consequence of the other principles? Most philosophical reviews of quantum physics have favored the first alternative, the one chosen by Bohr and long considered to be the only possible one. It is clear that the philosophical consequences would be radically different if the second possibility turned out to be the correct one. However, the dilemma is not one for philosophy but for physics to solve, since the second alternative is really a problem in theoretical physics having either a positive or a negative answer.

Science demands time, and philosophy calls for even more time, despite the impatience and the eagerness of the mind. Bohr's stipulations would have been a perfect example of wisdom had they been presented for what they are: practical rules for the physicist and necessary rules of caution to guide thought, even if they were to be only temporary. It is a pity that he went beyond this, for otherwise he would have gone down in history as the paradigm of a wise man, and not only—by no means a small achievement—as a truly great physicist.

PART THREE

FROM FORMAL BACK TO VISUAL:

THE QUANTUM CASE

✛

We HAVE just appraised the extent of the invasion of science by formalism. It is a disappointing realization, at least at first sight, and it may appear to bode us ill if our aspirations are of a philosophical nature—in other words, if we expect to understand. Who would pretend that his or her understanding is enhanced by surrendering to the language of signs, to a ghastly logic, offering nothing we can *see*, no source of light? One might be inclined to declare that we have touched the bottom, the incomprehensible, the cold foundations of the world. How could we be surprised, then, if so many curious minds, overcome with discouragement, turn away from the obscurity of science?

Where are we, anyway? We have been forced to give up a good deal of our intuition and our familiar language, which could no longer be trusted. Part of our representation of the world has been banned, and whatever remains is a world of atoms governed by signs, just as mathematics is, that collection of signs with multiple interpretations. We have suffered a very heavy loss but the gain is no less substantial. Thanks to science we have had access to laws, to a framework of the universe that is here to stay, and whose pure forms, repulsive as they may seem, are strong enough to inspire a new hope.

The laws, their consistency. And if we had lost nothing but gained everything? Imagine we could rebuild it all, recover our initial vision and our contented intuition of a world in which nothing was foreign. For this to be possible it would be enough to move a little closer to consistency, in order to realize that the visible world is in no way the Maya of Hindu philosophy, the universal illusion. It would suffice to show that our simplest vision and our humble, very ordinary language are the natural products of the laws, and that we therefore have the right to trust them.

It is to such a reconstruction that I now invite you. The task is an ambitious one, because it leads to nothing less than a reversal of philosophy's traditional approach. Instead of beginning with the reality of the world, jumping to conclusions about its principles on the basis of hasty remarks and frail generalizations, we must travel in the opposite direction, from the top down, as it were. Starting

161

with the laws so painfully conquered, we must descend back to the initial evidence, reconstructing it and justifying it along the way. If such a coherent vision is possible, if the intuitive and the formal aspects can really coexist, then it is irrelevant which one we take as our starting point, for coherence is a circle without a prescribed beginning or end. I, human, can enter the world through either sight or signs, it makes no difference.

Still recently, physicists have been engaged in such a task regarding the aspects that concern them. They have attacked that formidable fortress of formalism, quantum physics, whose principles, forged in the utmost abstraction, are the farthest removed from the transparence of reality as we perceive it. They have shown that it is nevertheless from those principles that such a transparence comes.[1] We shall now follow their trail. The path is hardly an easy one, for it goes through an area of science filled with snares, a true minefield. I hope, however, that it has been sufficiently cleared.

Some readers may wonder why, again, quantum mechanics. I would ask them to see, in what follows, only one example of a novel approach, which appears to open possibilities so far unexplored. These same ideas used elsewhere, enriched by having been successfully applied, might become still more effective, more convincing. That is a task that remains to be done.

I am well aware that the results that support my arguments are vulnerable, as all human enterprises are, and that they remain exposed to an eventual refutation in the light of new discoveries. However, whatever the future may bring, the path I have traced out could be rebuilt, and the method applied again. Ultimately, it is this method, this new philosophical tool, that matters the most.

---

[1] These works, due in particular to the physicists Murray Gell-Mann, Robert Griffiths, James Hartle, and the author, are presented in detail in the book *The Interpretation of Quantum Mechanics* by R. Omnès (Princeton, N.J.: Princeton University Press, 1994). They are also presented, together with recent results and improvements, in another, less technical book for physicists by the author: *Understanding Quantum Mechanics* (Princeton, N.J.: Princeton University Press, 1999).

Between Logic and Physics

THE OUTLINE OF A PROGRAM

NOTHING COULD BE more arid than the principles of quantum mechanics. Its concepts and laws are cast in a blunt, inescapable mathematical form, without a trace of anything intuitive, a total absence of the obviousness we see in the things around us. And yet, this theory penetrates reality to a depth our senses cannot take us. Its laws are universal, and they rule over the world of objects so familiar to us. We, who inhabit this world, cannot make our own vision prevail over those arrogant laws, whose concepts seem to flow from an order higher than the one inspired by the things we can touch, see, and say with ordinary words.

It is inevitable that such a vast theory should overturn the philosophy of knowledge's traditional assumptions. Surely, we can always assume, as Hume did, that our intuitive representation of the world is the direct consequence of our perception of reality. On the other hand, the wall that prevented us from understanding, and which Hume thought to be indestructible, is now almost in ruins. Hume did not believe that humans could ever know why there is so much order in the world, an order that we can see and speak about. Today we are faced with the opposite problem. We finally have access to the hidden order that governs the things we see and gives language its meaning. The experimental path advocated by Bacon has taken us much closer to the heart and the essence of things. We need not shut ourselves with Kant in the prison of innate ideas either; they can only restrict thought, while the arrival of formalism propels thought toward a future of unlimited possibilities.

That being said, we must admit that our inner self is torn apart by the opposition between formalism in our mind and concreteness before our eyes. If we really wish to understand, our first task must be to come to terms with this opposition. We need to conquer formal science, lift the new interdictions imposed by Bohr, the case of quantum mechanics being a good example.

There can be no doubt that the principles of quantum mechanics clash with common sense. We had better accept it up front rather than seek at all costs some artificial compromise. But such a recognition should not be a pretext for ruling out common sense as worthless, if only because we cannot do without it. Science is above all a product of experimenting, and an experiment is an action, even if it is guided by thought. Setting up a voltmeter is an action, as are installing a radioactive source and a Geiger counter, and moving the counter from one place to another. How could I describe all these actions other than by using ordinary language? Certainly not by talking about the voltmeter wave function. No one would contemplate saying, "Set up the voltmeter so as to give it such and such a wave function." It would be as inconceivable as imagining a driving instructor telling the student what must be done to the wave function of the brakes when the wave function of the photons from a traffic light has a certain form. Nobody would be safe in the streets any more. Giving instructions or directions, thinking about one's actions, communicating what we have seen, . . ., in short, everything pertaining to *practice* also belongs to common sense. And we have only considered science, while, more generally, the countless actions that are part of everyday life can only be represented in the familiar, ordinary way.

But the logic of common sense cannot handle events taking place on the atomic scale. Those events are governed by an altogether different physics, a universal physics, more general and extensive than the one ruling the world we can "see." Classical physics, the one familiar to our intuition, is only an extreme form that quantum physics adopts when it is applied on our scale.

Hence, if we really wish to understand, we must rely on what is known to be universal and founded on facts, rather than on what has already proved fallible. In physical terms, this means that we must not begin with the classical but with the quantum world, and *deduce* the former together with all its appearances. This deduction cannot simply result in the recovery of some fragment of classical dynamics; it must also be able to establish how and why common sense (that is, ordinary logic) can explain it. There lies the originality of our approach: *to deduce common sense from the quantum premises, including its limits—that is, to demonstrate also under which conditions common sense is valid and what is its margin of error.* We are well aware that our approach turns upside down the traditional explanation process dating back to the Greeks: we no longer explain reality from our mental representation of it, taken for granted without question; but it is this representation, with the intuition and the common sense that go with it, that we want to *explain*, beginning with our knowledge of the laws of the universe that science has revealed to us.

THE LOGIC OF COMMON SENSE

Amid the strangeness of the quantum world we may feel as lost as Alice in Wonderland, pondering which path to follow and which magic to trust. But Alice's father, Lewis Carroll, was a shrewd logician who secretly guided her. After all, isn't logic the best beam for those who have lost their way? So why not turn to it for help in trying to make some sense of this confusion? As we have seen, logic can be applied to any subject, provided we can clearly define three essential elements. First, we need to specify what we are talking about, a field of propositions, a universe of discourse, or, in other words, a domain of thought (*Denkbereich*). The second element provides the reasoning tools: the operations on propositions (not, and, or) and the relations of logical equivalence and implication (if . . . then). The third component of logic is a criterion permitting us to decide whether a given proposition is true.

Before applying these universal tools to the elucidation of the quantum world, we are going to attenuate their abstract character by fleshing them out a little. We shall begin by using them to discuss Newton's mechanics, a domain not too far removed from common sense, and therefore capable of shedding some light on its nature. Actually, we shall travel halfway through the distance separating the formal quantum world from common sense by giving the latter a touch of formalism.

For the time being, we shall consider only those propositions concerning the position and velocity of a physical object at some instant of time. They form the domain of kinematics, that branch of classical mechanics that precedes dynamics (in the natural order), the study of motion independently of its cause.

Our example will be a pendulum that oscillates in a vertical plane. We need one number to determine its position, the angle x formed

CHAPTER IX

by the string with the vertical, and another number v to denote its velocity. In this situation, the simplest proposition merely states the values (x, v) of the two quantities: "The position coordinate is the number x and the velocity is the number v." Notice that this presupposes that both numbers x and v can be exactly known, with infinitely many decimal places, if necessary. Now, such a proposition cannot describe an empirical fact, if only because it is impossible in practice to achieve infinite precision due to the limitations of measuring instruments. What's more, the above proposition is a priori incompatible with quantum mechanics in view of Heisenberg's uncertainty relations. And so, by whatever angle we attack the question, we have to resort to other propositions that better reflect experimental reality and are compatible with the quantum laws, known to be the most fundamental.

Suppose we measure the pendulum's initial position using an instrument with an accuracy of one second of arc and find that x is, say, 1,123 seconds. Given the instrument's precision capability, we can only be sure that x is greater than 1,122 and smaller than 1,124 seconds, and say that x equals 1,123 with an error Δx not exceeding 1. Similarly, given the possible vibrations, the effect of air currents, the shaking of our hand when we release the pendulum, or what have you, we will say that the initial velocity v is 0, with an error Δv not exceeding 0.01 mm/sec. The corresponding empirical proposition then is "The pendulum's position is between 1,122 and 1,124, and its velocity is between −0.01 and +0.01."

If we represent the situation graphically in a Cartesian coordinate system, the points (x, v) for which the above proposition is true are all inside a rectangle (the intersection of two strips parallel to the axes, such that the x-coordinate falls between 1,122 and 1,124 and the v-coordinate between −0.01 and +0.01). In this way, each elementary proposition in kinematics is unambiguously associated with a certain rectangle in the plane. The advantage of using a graphical representation is the ease with which the logical operations can be described when applied to regions of the plane or the space. We begin by considering more general propositions, referring to arbitrary regions of the plane instead of simple rectangles (fig. 1). To a given region D of the plane we associate the proposition: "The numbers (x, v) are the coordinates of a point in the region D." This proposition is denoted by D. Then, the proposi-

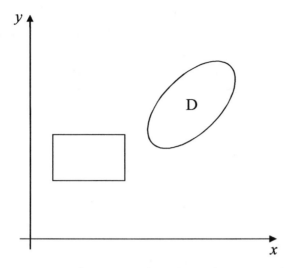

Figure 1. The configuration of an object (here, a pendulum) is defined in classical physics by a position coordinate x and a velocity coordinate v. When these quantities are given within some specified error interval, the corresponding point lies inside a rectangle. More generally, the point may lie inside a region D such as the one illustrated here.

tion not-D corresponds to the region outside D (the complement $\bar{D}$ of the set D); the proposition "D and D'" corresponds to the intersection of the regions D and D', and the proposition "D or D'" to their union.

It is also very easy to translate graphically the logical relations of equivalence and implication. D and D' are equivalent if the corresponding regions D and D' coincide, and the implication $D \Rightarrow D'$ corresponds to the case where the region D is included in the region D'. The significance of the above conventions lies in the fact that they satisfy the fundamental axioms of logic, as we know thanks to Boole. And so the logical operations become geometrical manipulations of sets, and the whole of kinematics, including the way we speak about it (that is, its logic), is reduced to a simple mathematical form.

To complete the application of logic to this particular branch of science we must specify the criterion by which the truth of a proposition is established. A proposition about reality is true precisely when it is in agreement with what actually is. Or, to employ

Tarski's elegant formula, "The rose is red" is true when the rose is red, in other words, when we can verify that the color of the rose in question is indeed red. In the case of kinematics, it is by actually measuring the pendulum's position and velocity that we can confirm the truth of a proposition of the kind we have introduced.

CLASSICAL DYNAMICS AND DETERMINISM

We can deepen our understanding of the logic of common sense by going from kinematics to classical dynamics. This new phase is not a mere repetition of the previous one, for it will shed some light on the nature of determinism and on the notion of truth as we usually understand it.

Formally, an elementary proposition in dynamics is simply a proposition in kinematics in which time is explicitly mentioned, for instance, "The position and velocity coordinates are in a certain region D at time t." To translate it into geometrical terms we now need a three-dimensional coordinate system in which to represent the points (x, v, t), but we shall not elaborate on the details.

The introduction of time brings in more than just another dimension, for the classical equations of motion make it possible to relate situations taking place at different times. If the kinematic coordinates x, v of the pendulum are assumed known at time t, classical dynamics allows us to deduce the new coordinates x', v' at some other instant (t') by solving Newton's equations. This relationship works in both directions of time, forward and backward (if we ignore the effect of friction), the instant t' coming after or before t.

In any realistic situation, when infinitely precise coordinates are not envisioned, we may consider the proposition a according to which the coordinates x, v are in a certain region D at time t. Each point x, v is transformed by the motion into another point x', v' at time t'. We shall denote by D' the region generated by the points x', v' as the point x, v describes D, and by b the proposition stating that the kinematic coordinates are in the region D' at time t'. It should then be obvious that the two propositions a and b are logically equivalent (fig. 2).

This reveals a purely logical facet of determinism: *Classical determinism is a logical equivalence between two propositions of*

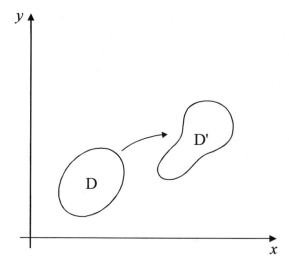

Figure 2. The logical nature of determinism. It is logically
equivalent to say that the position and velocity coordinates
are in a certain region D at a given time, and to say that they
are in another region D' at some other instant of time,
where D has been transformed into D' by the classical
(that is, Newtonian) laws of motion.

Newtonian dynamics with respect to two different instants of time.
Even if this remark may appear trivial, it nonetheless adequately
expresses determinism's main idea, according to which the past
completely determines the present; conversely, the present determines the past (in the absence of friction). Such was Laplace's conception of determinism.

The notion of truth also finds itself enhanced when time is taken
into consideration. We may witness an event taking place at time
t, and as such, it is indubitable. For example, I can see that the rose
is red at noon today, September 7th. Can I affirm that the deterministic consequence of this true proposition is equally true, as a
logical consequence, and that the proposition—which can only be
verified at a later time— that "The rose is wilted on October 15th
at noon" is also true? Two conditions must be met first. The most
important of the two by far is the existence of a universal law of
nature, be it theoretical or empirical, according to which all roses
wilt within one month. It is the type of law embodied in Newton's
principles in the case of dynamics, but it is clear that common sense

presupposes a great number of other, implicit rules. It is not scientific wisdom but plain common sense that prevents a man in love from cutting flowers one month ahead of Saint Valentine's day. The second condition for the inferred proposition to be true is merely a logical one: The above universal law should actually imply the truth of the proposition "The rose is wilted on October 15th at noon" from the truth of the initial proposition "The rose is red at noon on September 7th."

The above discussion may have appeared too simplistic, but it was only intended to bring us closer to a vision of the world where formalism and the reality of things are intimately connected.

WITH THE HELP OF AN ANGEL

The real difficulties begin when we enter the quantum world. If we wish to carry out to its conclusion the program outlined at the beginning of part three, we must be willing to give up almost all our old habits of thought, despite the fact that common sense is so well entrenched in our mind that it is practically impossible to ignore it, even for a moment. Nonetheless, it is imperative to assume at the outset only the formal principles of the physics we consider the most certain and the deepest, if common sense is to reappear in the end. It is the exercise to which we must surrender if we wish to be convinced of this wonder: the complete agreement between thought and reality.

If such is our human nature that it prevents us from overcoming our thought schemes, we can always imagine a new being exempted from our terrestrial limitations, a being capable of breathing the ether of pure theory, from which she would draw her sole inspiration, making her a kind of angel. Indeed, why not call upon an angel to help us out? This would be only a rhetorical device, similar to Voltaire's naive Huron, who represented simplemindedness in its purest form. Likewise, we shall find it useful to summon from time to time such a character, who would play the part of an expert in the formal approach, and be particularly suspicious of arguments based on all-too-familiar, but misleading, appearances.

Let us then imagine a newborn angel living outside the material world, in the realm of pure thought, where all her training will take place. Since her final destination is the earth, we must teach her all about the terrestrial world, but without yet showing it to her, in order to spare her too big a shock. Step by step, she must learn how we humans conceive this world. Our angel has a solid knowledge of logic and mathematics, for these disciplines are easy to master in her paradise. We must first explain to her, gently and clearly, what matter is.

She begins then by learning the fundamental laws of nature, in particular those of quantum physics: the world is made up of particles described by wave functions that evolve according to Schrödinger's equation. Since the angel must also learn the necessary theoretical background, we must take her down the path that led humans from common sense to the logical formalism of classical physics, but in the opposite direction, that is, going from the complete mathematical formalism to propositions that humans can understand but that are also perfectly valid and clear for the angel, who is a pure theoretician.

There is no simpler proposition in physics than one that states the value of some physical quantity at a given instant of time. However, such a proposition is not as elementary as it might appear, for the notion of physical quantity (a position or a velocity coordinate, or energy, for instance) is surprisingly abstract in quantum mechanics. Without going into the details, let us just say that a physical quantity (also called an *observable**) is mathematically expressed by an operator, that is, by a kind of device that manufactures functions: given a function as input, it produces as output another function. The formal character of such a concept is particularly manifest when the physical quantity is a velocity component, for in this case the corresponding operator calculates the derivative of the wave function and divides it by the pure imaginary number i. It is then better to leave it at that, and to notice that the angel sees no problem here, for she has no idea of what it means to have a measuring instrument that would give a concrete interpretation of such a quantity. For her, everything is perfectly clear, for everything is purely abstract, purely mathematical.

What is important for her is that the proposition "The value of the physical quantity A is in the interval D at time t" is absolutely clear. This is what we shall call a *property,** expressing, for example, the fact that the angle x formed by our pendulum with the vertical is between 1,123 and 1,124 seconds at some particular time. But what is the angel to make of such a property? She does

171

not have the slightest idea of what a "pendulum" might be (other than a collection of atoms) or of the intuitive meaning of "position" (which is for her merely a certain operator). Since she does not know anything about reality yet, we cannot speak of measurements obtained by using an instrument which, for us, would give some concrete meaning to the above property. The notion of "concrete" does not belong to pure theory, and a measuring instrument would only be for the angel a large quantum system of no particular interest. For her, nothing of all that is relevant, only this, which is cast in formal terms, the only language she understands: The statement of the property uniquely determines a certain mathematical object that completely characterizes it.

In short, we are still in the formal domain, and what is for us a property that we can understand has a meaning for the angel only because the mathematical dictionary of the theory offers her a perfect translation. Such a mathematical translation of a property is called a *projector** (or projection operator). Let us just say that it is also an operator, whose only possible values are exactly 0 or exactly 1.

We shall give only one example, in the simple case of a property stating that the position coordinate x of a particle is between 2 and 3 (with respect to a certain unit of length that it is not necessary to specify). The corresponding projection operator, which is only meaningful when applied to a wave function $\Psi(x)$, may be seen as an operation whose effect is, so to speak, to clip the wings of $\Psi(x)$: the function remains unchanged for values of x greater than 2 and smaller than 3, while under the action of the operator all values of $\Psi(x)$ become 0 outside this interval.

We can also see in this example in what sense the projection operator can take on the values 0 or 1. If the input function is different from 0 only inside the interval from 2 to 3, then the "wing-clipping" operation will leave it unchanged—it is multiplied by 1; if the input function is already 0 inside this interval, it will be 0 everywhere after the operator is applied—it is multiplied by 0. To try an explanation beyond this particularly simple case would take us too far. The essential thing to remember is that each property has a corresponding operator whose only possible values are 0 or 1.

Zero or one only! Remember our journey through formal mathematics, where the symbols 0 and 1 were used to denote false and true. We have just met some mathematical objects, the projectors, which represent properties and cannot take on values other than 0 and 1. Could this mean that a property can only be true or false, as we all naively believe, the same way we believe in the principle of the excluded middle? Let us promptly add that all this is merely a rough indication, a quick glance through a sudden clearing in the fog, and that we still have a long way to go before we can be out of the wood. We are at least heading in the right direction, and the angel is beginning to speak in a tongue that resembles our own language.

OBSERVABLES

I do not know whether you, dear reader, feel like being an angel. Maybe sometimes you do but, alas, not me. However, we both know that looking at the same question from different angles may help to understand it. Let us therefore take a look at the tricky notion of an observable from a more human point of view.

You may be familiar with the notion of a random variable in classical probability theory. If you are not, here is essentially what it means: There is an object, a die, say. It has six faces which we label from a to f. When the die lies on a table showing the upper face a, we say that this is an event, which can also be denoted by a. A number is associated with each event. For instance, face c is painted with three dots, and the corresponding number is 3. Finally, each possible event is assigned a probability, which depends on physical conditions (whether the die is loaded or not or how it is thrown). A collection of the three notions, "events, numbers, probabilities," is called a random variable. It is simple and useful, and you may rely on it to compute how much you can expect to lose in your next visit to Las Vegas.

Suppose now that your die is a quantum die. There might still be six different events and the same numbers as before associated with them. Let us leave the probabilities aside—they depend on how the die is thrown, its state or initial wave function, if you prefer—and rather introduce a new notion that did not occur in the case of the classical random variable: Each event is expressed by a statement, for instance, "face a is uppermost" or "event a occurs." This is a sort of an intrusion of logic into the game, each statement receiving one of the two truth values, 1 or 0, according to whether it is true or false. We must explicitly require that when statement a is true, all others should be false.

A simple mathematical way of realizing these properties of truth values is to represent every statement essentially as the angel did, that is, by a projector. The logical content of a statement is much better expressed in that way than in plain words. When considering, for instance, the value x of a particle's position coordinate, the two statements "x is between -1 and $+1$" and "x^2 is between 0 and 1" have different wordings though the same meaning. Identical projectors are associated with them, but we have already mentioned this.

An observable, a quantum physical quantity, thus appears as having both a logical content and a quantitative aspect. It is a collection "numbers, statements" or, in mathematical terms, "numbers, projectors." After all, one should not be surprised that some statements, the statements of some occurrence, should be an intrinsic part of a physical quantity in a purely probabilistic theory. Once we realize this, quantum observables lose much of their mystery. An observable, as usually meant, is obtained by multiplying each number (each possible value of the observable quantity) by the corresponding projector, and then adding the results. Therefore, the observable contains the information on the possible values of the quantity and the corresponding statements, as expressed by the projectors.

Mathematicians call the numbers "eigenvalues," the projectors are said to project on "eigenvectors," and the sum we have just mentioned is called von Neumann's spectral theorem. Frankly, not a very transparent language. But after all, mathematics never provides a meaning. It rather conceals it, and we have just seen that meaning must come from inside physics itself, from a specific way of describing nature.

RUDIMENTS OF A QUANTUM DIALECT

At present, our angel is not very loquacious. She knows only how to repeat sentences such as: "The value of the physical quantity A is in the region D at time t," for various A, D, and t. She does not even seem capable of the most elementary logic, to say, for instance, "The value of the physical quantity A is in the region D at time t *and* the value of the physical quantity B is in the region Δ at time t." She is not yet what one might call a brilliant talker.

There is an essential difference between her language and ours, concerning the notion of commutativity. The noncommutativity of operators plays a crucial role in quantum mechanics. It consists in the fact that the product of two quantum physical quantities A and B is not in general commutative, that is, the product AB is not equal to the product BA. Remember that a physical quantity A is an operator, roughly equivalent to a computer program, which transforms any given wave function Ψ into another function $A\Psi$. The physical quantity B is associated with another program which, applied to the function $A\Psi$ produces the new function $BA\Psi$. If we had applied program B first, followed by program A, we would have obtained a function $AB\Psi$ which need not be equal to $BA\Psi$.

We can put the above explanation in more concrete terms. Since a wave function is associated with the state of a physical system, let us take as our system a potato and some water. Instead of operators associated with physical quantities A and B, we shall consider so-called dynamic evolution operators, which exhibit the same lack of commutativity as the previous ones. Let A be the evolution operator that transforms the wave function F of a raw potato into that of a boiled potato, and let B represent the evolution operator transforming a whole potato into a crushed one. If we perform the operations A and then B, in that order, we obtain $BA\Psi$, the wave function of a delicious mashed potato, and we can throw away the boiled water. On the other hand, if we follow the reverse order AB, that is, we crush the potato before boiling it, we end up with an unsavory liquified potato. Cooks unanimously agree, as do angels and mathematicians, that operations do not necessarily commute.

However, some physical quantities do commute. This is the case of the X and Y position coordinates of a particle, that is, its coordinates along the first and second axes, respectively. In this case it is possible to combine the corresponding properties using "not," "and," "or" in a perfectly consistent fashion and to say, regarding for example a hydrogen atom, that "the value of the X coordinate of the electron's position lies between 0.7 and 0.8 Å *and* that of its Y coordinate lies between −1.1 and −1.2," for there is a projector for which this property is meaningful. Using such sentences as

175

elements and combining them by means of "and," "or," "not," we can describe positions having an arbitrary geometry. The elementary properties X and Y describe all rectangles in the plane and, as every printer knows, any plane figure may be drawn using rectangles of sufficiently small size.

Things are completely different if we try to combine a position coordinate, X, say, and the component U of the velocity in the same direction, because the corresponding projectors do not commute. Hence, a proposition stating that "the value of X lies in the interval D *and* that of U in the interval $\varDelta$" is meaningless for the angel, for it is impossible to associate a projector with it. *Therefore, in quantum mechanics there are propositions that can be expressed in ordinary language but which have no meaning due to the underlying formalism.*

John von Neumann, who established the basis of quantum logic, was particularly struck by this interdiction, and sought, together with George David Birkhoff (1884–1944), a way to describe propositions involving quantities that do not commute. They partially succeeded, by creating a projector with a vague connection to the contested propositon: "The value of X is in D *and* that of U is in $\varDelta$." Unfortunately, the connection between the usual meaning of the sentence and the projector became extremely weak, for while the proposition determined the projector, the latter no longer allowed one to recover the proposition. There was no correct dictionary any more, no longer a language but only a babble, and a nonsensical one at that, because it did not obey all the rules of logic. Birkhoff and von Neumann then wondered whether the logic of the quantum world did not after all obey laws that were different from—and less restrictive than—the sacred ones of Aristotelian logic.

It does not seem possible to adopt such an intrepid idea, first of all, because, as we have just seen, the mathematical translation of ordinary language is no longer faithful. The second reason is a matter of coherence: the theory's underlying formalism is mathematical, hence Aristotelian; on top of it we would now need a non-Aristotelian construction to interpret the theory, in other words, to reconcile empirical physics with common sense, the latter being as Aristotelian as it gets. The quantum non-Aristotelian component would then be flanked by the mathematical (Aristotelian) and the empirical (also Aristotelian) ones. A rather indigestible "sandwich" to swallow, leading us to the third and last reason for rejecting the above approach: more than fifty years later, it is hardly more developed than it was on the first day. We shall therefore abandon it, to insist on the advantages of sticking to more conventional—more rational, in fact—logical forms.

HISTORIES

A novel idea, simple and fruitful, was introduced in 1984 by the American physicist Robert Griffiths, from Carnegie Mellon University. Instead of looking only at isolated properties taking place at a single instant, Griffiths proposed considering what amounts to the history of a physical system, that is, a sequence of properties occurring at successive instants. Despite its simplicity, the idea had not been explored before because it appeared to be incompatible with the noncommutativity of operators, the same difficulty von Neumann had encountered. It was thought to be impossible to be able to say of an electron visiting the Louvre Museum: "At 9:12, it is in the Antiques Room; at 11:30, its velocity is between 2.3 and 4.5 km/h; and at 12:47 it is in the Corot Room."

A history is simply that: a sequence of various properties taking place at different times. Each property expresses the fact that the value of some physical quantity is in some region (of values) at some instant of time, and the history merely lists them, the choice of the physical quantities, the regions of values, and the instants of time being more or less arbitrary.

Compared to a single property, a history has a much greater potential to describe what goes on. We may say that a history is to an isolated property what a film is to a snapshot. It turns out that Griffiths' histories may serve as a language to describe all of physics, a kind of universal language that allows us to speak of all physical events without exception.

Histories should not appear all that mysterious, because we have always used them to describe experiments and other situations in physics. Here's an example of this important point, and it would be easy to find many others. A physicist tells another about one of his experiments: "A neutron comes out of a nuclear reactor through an opening in the armored wall. It then traverses a silicon crystal; it comes out (after diffraction) with a velocity that depends on its direction; the velocity is then selected by forcing the neutron to pass through a narrow window; next, it hits a nucleus as it travels through a uranium block; the collision results in the fission of the nucleus, which breaks up into several pieces; one of these is a xenon nucleus which, finally, enters a counter's detection zone."

That is a history, which a theoretician could rewrite using projectors to put it in Griffiths' standard form, after having specified the time of each event. In this form the angel understands it as clearly as we do.

THE ROLE OF PROBABILITIES

The reader must have noticed that quantum probabilities have not yet been introduced in the initiation to the world our angel is going through. This may seem surprising, given their predominant role. We shall now see them enter the picture in a most unexpected way, not as a measure of chance but as a tool to complete the logic and give it a consistent meaning. It is in fact thanks to probabilities that we can convince ourselves that some histories are meaningful and others do not make sense, and, among those that do, define logical equivalence and implication in the quantum world. Probabilities thus lie at the very heart of the theory, and their role goes far beyond the mere description of chance.

This is a consequence of their formal mathematical structure, known and accepted by the angel, and which we should take as a mathematician would, without the slightest regard for practical applications. For the mathematician, probabilities are simply numbers assigned to events (which are properties or histories, in our case). Those events form a complete family (they are mutually exclusive and cover all possibilities). Probabilities are subjected to only three conditions: they are positive numbers, they can be added if two events are mutually exclusive (this is called the additivity condition), and their overall sum is equal to 1.

To add some substance to his histories, Griffiths assumed that each of them had a certain probability. Such an assumption appears quite reasonable in the case of the nuclear experiment described above. Indeed, the neutron might not have come out of the reactor, or might have missed the window or failed to traverse the uranium block, or, even if it did, no reaction might have taken place. Griffiths proposed an explicit mathematical formula to calculate the probability of a given history, a formula that was later seen to follow from some simple logical considerations.

He then observed that the additivity condition for probabilities significantly restricted the class of all conceivable histories. This condition is expressed by a mathematical equation involving the projectors of the various properties occurring in the history, an explicit equation that can be checked by computation. Griffiths then called consistent those histories that satisfy the additivity condition. For example, the history associated with our nuclear experiment belongs to a consistent family and has a well-defined probability, perfectly satisfactory from a mathematical point of view.

To help the reader understand the notion of a consistent history we shall present a most astonishing counterexample, arising from interference phenomena, in which probabilities are not additive. It goes like this: A photon comes out of an interferometer and hits a screen. We can describe the hit by imagining the screen subdivided into small regions (for example, each region may consist of a single grain of photographic emulsion). In this way we have as many different histories as there are screen regions. Nothing prevents us from assigning a probability to each of these histories. These probabilities turn out to be additive and perfectly acceptable. Their values, computed according to the theory, clearly reveal the presence of interference fringes.

Things become both more subtle and more interesting if we try to determine the photon's path before it hits the screen. We may choose an instant when the wave function derived from Schrödinger's equation is made up of two parts, each one localized in a different arm of the interferometer. We now enrich the previous histories by specifing that the photon is in one or the other arm at this particular time. At this point, the truly dramatic interest of Griffiths' additivity conditions reveals itself, because it is impossible to satisfy them. No additivity, no probabilities, no meaning. Hence, the statement that the photon followed one path rather than the other is meaningless, despite our habits of thought, which cry out that something is amiss here.

This is a remarkable result, because it suggests that certain histories are meaningful but not others, at least if we agree that only histories to which we can assign a probability have meaning. But what meaning? This is what we are going to see next.

THE LOGIC OF THE QUANTUM WORLD

The most important quality of Griffiths' construction is to endow quantum physics with a logical structure of its own, as I have tried to demonstrate. It is precisely this feature that allows us to go from a purely formal theory to something we can talk about using ordinary words and, above all, something we can reason about rigorously and naturally. As we already know, rigorous reasoning requires a logic that is complete and sound, starting with a domain of propositions where we can say "not," "and," "or," "if . . . , then."

A domain of propositions suitable for describing a quantum system involves a family of histories verifying Griffiths' additivity conditions (a consistent family), so that probabilities may be assigned to histories. The logical operations "not," "and," "or" are then quite obvious, as when we say, "The neutron was *not* there at that instant," "It could have traveled this way *or* that way," "It went through a channel in the reactor's wall *and* then through the window." The last condition for a sound logic will be fulfilled if we can find a satisfactory definiton of logical implication, that fateful "if . . . , then," logic's keystone, without which reasoning is impossible.

Here's how we may introduce it: In our consistent family of histories, probabilities are irreproachable from a mathematical point of view and, in particular, they are additive. We can then borrow from classical probability theory the notion of conditional probability. This is, by definition, the probability of an event b occurring assuming that another event a has already occurred. This was what Don Juan had in mind when he wondered, what is the probability that the next girl I meet be blond, assuming, of course, that she is pretty? Such a conditional probability, denoted $p(b \mid a)$, is defined mathematically as the quotient of two probabilities: that of a and b occurring together (the next girl is both blond and pretty) divided by the probability of a (the next girl is pretty, regardless of the color of her hair). We shall say in quantum logic that a implies b if the conditional probability $p(b \mid a)$ is equal to 1. At the risk of repeating myself, let me insist: the introduction of probabilities makes inference, hence *reasoning*, possible.

The logical equivalence of two propositions a and b is now an immediate corollary, for it amounts to having both a implies b and also b implies a. Here is a nontrivial example of a logical equivalence, even if we are jumping the gun here by referring to measurement theory, to be introduced later. If a photon detector is placed behind a circular light polarizer, we shall show that the proposition expressing that the detector has recorded (an empirical proposition involving only the detector) is logically equivalent to another proposition expressing the value of a component of the photon's *spin** (that is, indicating how the photon spins), which is a proposition referring only to the microscopic world of photons.

These logical conventions are not arbitrary, provided they are applied only to consistent histories (those having "good" probabilities, that is, satisfying Griffiths' conditions). The axioms of logic, those of Aristotle and Chrysippus, and such as they were later formalized by Boole and Frege, are then completely satisfied. And so the major obstacle encountered by Birkhoff and von Neumann in dealing with logical issues has now been removed, and Aristotle is back. Common sense should follow shortly.

COMPLEMENTARITY

Unfortunately, the return of a sensible logic does not entail that common sense is back too, for the quantum world is still full of subtleties. As a matter of fact, a given quantum system may be described by many different families of histories. In the nuclear experiment presented above, for example, we could specify at a given instant the velocity of the neutron instead of its position. Depending on our choice, we would have two different domains of propositions (or of histories), two different logics that cannot be embedded into some other larger and consistent logic.

Bohr was already aware of this peculiar fact, and he had even raised it to the level of a principle: complementarity, one of the pillars of quantum mechanics. In reality, he was referring to the knowledge acquired by using two different experimental devices. For instance, if a silicon crystal diffracts a neutron we should, according to Bohr, speak of the neutron as a wave, while its detection by a counter would force us to construe it as a particle. Quantum logic shows that such a multiplicity of representations is not just imposed by external instruments but truly intrinsic to the quantum realm, even when it remains unobserved. Besides, it is certainly not a new principle but a direct consequence of the logic, which is itself determined by the other laws.

The existence of different logical frameworks in which to speak of the same object is nothing new, and we have already seen examples of this in connection with monogamous and polygamous marriages. It was precisely such a multiplicity of universes of discourse that prompted logicians to introduce the notion of a domain of propositions. What is new here is the fact that this sort of subtlety, usually of minor significance, now becomes essential for speaking of the material world, due to the quantum character of the laws.

The existence of thousands of possible ways of speaking of the same object, all mutually exclusive, might conceivably lead to internal contradictions or paradoxes. For example, what to do if, within a certain logic, hypothesis *a* implies conclusion *b*, while in another logic *a* implies not-*b*? We would then have no choice but to give up logic altogether—and curse it. Fortunately, it can be shown that in quantum physics such a tragic mishap can never happen, and that *a* implies *b* in every logic in which both propositions have a meaning: *there cannot be any paradox or internal contradiction in quantum mechanics.* This is an amazing result for a domain where it was long believed that paradoxes lurked at every corner. "The Lord may be subtle, but He is not wicked."

A LOGICAL LAW OF PHYSICS

Even if we are still in the abstract realm, certain signs are already pointing to a closer relationship with reality. The language of histories allows us to describe the quantum world and to *reason* about it, in terms that are as clear for us as they are for our angel. This language closely resembles the one a physicist employs when arguing about an experiment, with terms that are inspired by the most naive empiricism and based on visual representations. Thus, the physicist may say, "A xenon nucleus has reached the detector, *hence* there has been fission, *hence* the neutron has hit the uranium block, *hence* it went through the second window, *hence* I can determine its velocity, and so recover all the relevant data for a detailed study of the experiment."

This argument, despite its elementary character, is based on a logical implication that can now be completely justified every time the word "hence" is pronounced, and it reveals what is necessary for understanding physics. From our point of view the argument is intuitive, while it is absolutely rigorous for the angel or the mathematician. Each one of the statements following the unique hypothesis ("A xenon nucleus reached the detector") is supported by a calculus rooted on the fundamental principles: it would be enough to check Griffiths' conditions and compute the conditional probabilities that validate the logical implications. Thus, perhaps in an unexpected way, the drastic abstraction we had to accept in order to ensure complete logical consistency has led to a vision of the quantum world that is extremely close to the physicist's intuition. Let us add in passing that such was not the case with Bohr's traditional interpretation, which was incomparably more remote from intuition.

We shall now complete the list of fundamental principles of the theory by adding a new one, of a logical nature, that makes it possible to think the world, and not just compute it: *Every description of a physical system must consist in propositions belonging to a unique and consistent quantum logic. Every argument regarding the system must be supported by logical implications that can be demonstrated.* It should be clear that such a principle is rooted in the depths of physics and does not rely on the presence of any observer, which is totally fortuitous, if not irrelevant, contrary to what had long been believed.

Thanks to this new principle the angel can think like us, or rather better than us, because she knows perfectly well what it is permissible to think. As for us, this rule allows us to think in objective terms, without dreaming that logic is only a figment of our imagination. The two most important ideas to remember are first, that logic has its source in the laws of nature; secondly, that this logic of things cannot be dissociated from the existence of probabilities and, ultimately, from the necessary presence of chance. On this new foundation, built entirely from first principles, we shall now construct anew both common sense and the intuitive representation of the world.

Rediscovering Common Sense

Our next task will consist in nothing less than recovering the everyday vision of the world, as it is revealed to us by common sense and visual intuition. We shall start this time from the fundamental laws of nature, which are ultimately of a quantum, and hence formal, character. Since we accept being guided only by logical consistency, our approach will be entirely deductive. It will be based exclusively on the principles of quantum mechanics and, in particular, the logical principle stated at the end of the previous chapter will play a crucial role. What we shall accomplish will not be the mere recovery of common sense, but of something more refined and enlightening that will tell us precisely when common sense can be trusted and when we might be misguided by it, be it ever so slightly.

THE WORLD ON A LARGE SCALE

As long as we have decided to rely only on the principles of quantum theory, it will be convenient to seek once again the help of our familiar angel, who is still trying to figure out how we humans view the world.

The scale of this human world is very large when compared to the world of particles, and atoms appear to us extremely small, so small, in fact, that we cannot see them. We must therefore consider visible objects, accessible to our senses, macroscopic, that is, physical systems made up of large numbers of particles. Everything a human being can see or touch to forge his or her intuition belongs to this category: dust, trees, stones, and machines, all the way to the sun and beyond; in a word, the entire domain of classical physics.

Let us first remark that the notion of an object, so plain and primitive for common sense, is not at all obvious from the standpoint of quantum physics, and it is therefore puzzling for our angel. For her, a physical system is an assemblage of particles whose mutual interactions are known, usually a set of nuclei and electrons. If we consider from this perspective an ordinary object, an empty bottle, say, the quantum principles will only take into account the particles forming the bottle, and will therefore treat on an equal footing a multitude of different objects. This is due to the fact that the atoms that make up the bottle could, without changing their interactions, adopt thousands of shapes to form a thousand different objects: two smaller bottles, six wine glasses, or a chunk of melted glass. One could also separate the atoms according to their kind and end up with a pile of sand and another pile of salt. A rearrangement of the protons and electrons to transmute the atomic nuclei without modifying the nature of their interactions could also produce a rose in a gold cup. All these variants belong to the realm of the possible, of the multitude of forms that the wave functions of a given system of particles may take.

To be sure, objects can also be defined in quantum mechanics, and in fact each object corresponds to a certain class of wave functions that an indefatigable human computer could completely specify. Our angel can therefore learn this notion of an object, and she might even appreciate better than we do its fuzzy character (is a bottle containing two atoms still an empty bottle?). She is nevertheless still far from understanding the classical description of an object. For her, the positon of a pendulum or that of the hands of a clock is still at this stage a mathematical operator. The quantum nature of physical quantities has changed nothing, except that some of them, which we may call classical, have been singled out among the myriad of quantities describing the atoms, the ultimate components of matter and objects. In the physicist's jargon, the first observables that could become classical after analysis are called collective physical quantities, and the others are said to be microscopic. Thus, the positions of the pendulum or the clock hands are collective variables, as are all those occurring in classical physics. In like fashion, collective observables of velocity may be defined, but they do not commute with position coordinates. We have still a long way to go before rejoining Newton's or the engineer's concrete representation of things. But in order to do that and share their "candid" vision of the world, the angel must learn something else.

185

CHAPTER X

The Logic of Common Sense

The archangel who initiates the young angel into the secrets of the terrestrial world, our novice's final destination, begins by defining an object as a collection of wave functions. He then shows her how to obtain, from the first principles of the theory, the collective observables that describe the object (this is more than we humans can do at the present time, but research in this field is progressing rapidly). Thus, a pendulum becomes for the angel a metal ball (she is familiar with the quantum theory of metals) attached to a metal wire. The wave function indicates that the atoms form a ball of a certain radius, and something analogous specifies the shape of the wire.

The archangel then explains that humans prefer to focus on the coarsest features of objects, rather than take into account their rich internal structure. This is due to the imperfection of their senses. "They are wise to do so," replies the angel, "and I am willing to do likewise and only retain, of the pendulum's wave function, its dependence on the coordinates of the ball's center, and neglect all the rest."

At this point the archangel helps the angel through the decisive transition from the quantum to the classical world. Everything seems to separate these two visions of reality: on the one hand we have wave functions, physical quantities that are operators, a dynamics governed by Schrödinger's equation; on the other we have position and velocity variables that are ordinary numbers, and the dynamics is Newtonian. How to go from one to the other? It can be done but, we must admit, not without some powerful mathematical means. This is why mathematicians have developed, since the end of the 1960s, a whole new branch of analysis (known as microlocal analysis or pseudodifferential calculus) thanks to which an operator acting on the pendulum's wave functions can be associated with a function of the classical variables of position and velocity. Such a function is called the symbol of the operator. In this way a dictionary can be created, giving the translation of a great many quantum words into classical terms. With a bit of practice, the angel soon becomes familiar with this classical language.

The archangel may now explain to the angel what is a proposition of classical kinematics, as we have done in the previous chap-

186

ter (basically, it amounts to considering a cell in the space of the classical position and velocity coordinates). "I have no objection to talking about these things for fun," observes the angel, "but all that really means nothing. For you have told me that the properties permitted by the first principles are those that can be associated with a quantum projector, and what you are now telling me is quite different." The archangel then discloses to her a theorem showing how to associate a quantum projector with a classical region like the one indicated above, provided that the region is sufficiently large (compared to Planck's constant) and its border sufficiently regular.

"This is extraordinary," exclaims our angel, after having familiarized herself with the theorem and its consequences. "This result clearly shows that what you told me regarding classical propositions amounts to a quantum proposition. One must only be careful not to make statements too subtle to be expressible in classical terms, then one can speak both the quantum and the classical languages by translating the latter into the former. Is this the way humans think?"

"To tell the truth," replies the archangel, "their language is ordinarily less elaborated, but on the whole, yes, that's essentially how they perceive the world and talk about it."

The angel has yet to understand how to reconcile Schrödinger's quantum dynamics with Newton's classical mechanics. Here again, she must resort to mathematics in order to effect the necessary translations. In particular, she realizes that this relationship is only approximate, due to the fact that we are interested only in big objects, and without examining them too closely. Roughly speaking, the regions of the space of position and velocity coordinates get distorted during classical Newtonian motion. At the same time, the quantum projectors that express the corresponding properties evolve in a parallel fashion according to Schrödinger's equation. The correspondence between the region and the quantum property is nevertheless approximately preserved (the errors involved are well known). "But then," realizes the angel, "we can tell the history of a macroscopic object in the classical language without violating the fundamental quantum principles! I must practice this if I want to be fluent when I talk to humans."

The angel is now persuaded that a "correspondence" exists between classical and quantum properties. This correspondence is

187

preserved by the passage of time, at least in most interesting cases arising in practice, due to the harmonious agreement in the respective evolutions of classical and quantum dynamics. However, the correspondence is not an identity, and it is subject to certain conditions and prone to errors. The first result in this direction was obtained by the Dutch physicist Paul Ehrenfest in 1927. Bohr had previously introduced what he called the "correspondence principle," which expressed in terms still vague the expected agreement between the two dynamics. As is often the case in the history of quantum mechanics, the principle had preceded a theorem, and Bohr's correspondence principle has now been incorporated into the logical framework we have adopted thanks to more advanced mathematical methods.

In order to understand that we are in the presence of a correspondence and not of a perfect agreement, let us examine some of its limits. For example, it is not enough for an object to be large in order to exhibit a classical behavior. In particular, there exist objects whose motion is chaotic (atmospheric turbulence, for instance), and for which the correspondence is severely hampered. The motion of such systems results in a strong distortion of the classical cells, and consequently the correspondence between classical and quantum physics lasts only for a limited period of time. Nonetheless, the vast majority of objects either on earth or in the skies present a good correspondence between the fundamental laws of the quantum domain and the classical laws of the large-scale world in which we live.

Let us listen to the angel one last time: "I am elated," she says, "because I now understand how humans describe the world and also how they see it evolve. What a pleasure they would experience if, besides seeing, they also understood what they see."

"But they do understand!"

"What do you mean? You had assured me that the only sensible way to describe the world was in terms of the histories of a consistent quantum logic. You made that into a principle and convinced me that one can only reason by using logical implications which can be demonstrated. So far, you have only showed me how humans describe the world, but I don't see the connection with a quantum logic of consistent histories, nor how humans can reason in a sensible way, compatible with the first principles."

"Humans do that by using what they call common sense. It is a kind of logic well suited to the world they live in. The way they talk about their world and their mental representations of it are not completely rigorous, but it is nevertheless the same representation offered by the theory or, rather, it is a direct, *demonstrable* consequence of the theory, perfectly valid within its domain of application. The theory offers a purely formal representation, while humans use an empirical representation stemming directly from practice and experience.

"When humans see the objects around them, they estimate their position and velocity through the use of their senses. But these are not fine enough to appreciate the manifestation of quantum effects, so what humans perceive is perfectly expressible by classical propositions. To sum up, we may say that the mathematical inference of propositions in classical physics from quantum principles provides a faithful image of the way humans perceive the ordinary world and of their mental representation of it."

And the archangel goes on: "When human common sense reasons by saying 'if . . . , then,' what happens is this: humans mentally consider a cell in the space of position and velocity coordinates. This cell is estimated, however imperfectly, by their brain. They also instinctively (that is, out of habit) create a mental image of another cell that can be deduced from the first one by a Newtonian motion. They now say that, if the initial situation that corresponds to the first cell happens, then the situation corresponding to the second cell will take place after a certain time: they say that if an apple falls from a tree, it will hit the ground directly below. Of course, they reason likewise in many other situations as well, but the ones we consider here are at the origin of their vision of the physical world."

"I can see from what you say," observes the angel, "how humans reason using their common sense, and how they render it precise with the help of Newtonian physics. But I am not convinced that their reasoning is correct. The real laws of the world are quantum, and you have told me that only a consistent quantum logic permits us to describe the world and correctly argue about it. Now, common sense human logic is not at all of that kind; hence humans are only capable of faulty reasoning."

"Absolutely not! Their reasoning is sound. I have already told you how their evaluation of a given situation may be translated in terms of quantum projectors, and how the change in time of those projectors closely parallels the classical evolution of the situation. Using this correspondence it could be shown that common sense logic is actually also a logic of consistent quantum histories, and that common sense arguments are ultimately the verbalization of implications that can be demonstrated in quantum logic. This identification of common sense logic with a particular quantum logic is, of course, not perfect; there are exceptions, and its implications are only approximate. But the approximation is excellent in most cases. In other words, the probability for common sense to be wrong is practically always negligible, as long as it deals with macroscopic objects and does not approach too closely the world of the infinitely small."

"Thank you, master. Thanks to you I have understood how humans see their world and think about it in their own peculiar way, which is quite appropriate for the things they can immediately perceive. You have convinced me that their representation of that world and their common sense are perfectly legitimate—at least practically all the time and on a sufficiently large scale—even if the laws of reality are ultimately quantum and formal. I am now ready to go down to earth and meet those men and women you have taught me to respect. Haven't you told me that they have discovered the principles I have learned from you? Humans too are therefore aware that their ancestral modes of reasoning are the fruit of those laws."

DETERMINISM

A good example of the approach we have just presented is the way it clarifies the relationship between classical determinism and quantum probabilism. As we have already pointed out, determinism consists in a logical equivalence between classical propositions referring to two different instants of time. In the absence of friction, this equivalence holds in both directions: from the present to the future (the ordinary meaning of determinism), and also from the present to the past, which entails the possibility of recreating the past and is ultimately the basis for the existence of memory. Things are not so simple in the presence of friction, but we shall leave out this case.

The essential point, which has been understood only lately, is that classical determinism is a direct consequence of the quantum laws, despite the probabilistic nature of the latter. Reconciling these two seemingly incompatible points of view was only possible after classical determinism lost its absolute character and also ceased to be universal. Each of these two aspects is important, so it will be worth explaining them further.

Classical determinism is only approximate, as it is easy to see by some examples. Let us first consider an extreme case involving the earth's motion. What can be more deterministic than the fact that the sun rises every day? We know that the earth rotates around the sun according to Kepler's laws. This is a consequence of the principles of Newtonian dynamics and also, with a good approximation, of the principles of quantum mechanics. It is this notion of good approximation that we wish to make precise.

It is known that quantum mechanics allows for the existence of "tunnel effects" by which an object suddenly changes its state due to a quantum jump, something that would not be possible through a continuous classical transition. Many examples of such an effect are known in atomic and nuclear physics: it is precisely by a tunnel effect that uranium nuclei spontaneously decay, and two protons at the center of the sun may come close enough to start a nuclear reaction.

Even an object as large as the earth may be subject to a tunnel effect, at least in principle. While the sun's gravitational pull prevents the earth from moving away through a continuous motion, our planet could suddenly find itself rotating around Sirus through a tunnel effect. It would be a terrible blow for determinism. We went to bed the previous night expecting the sun to rise the next morning, only to wake up with a view of an even brighter star, which during the night gives way to unknown constellations.

A theory that permits such events to happen may well make us feel uncomfortable. Fortunately, even if determinism is not absolute, the probability of its violation is extremely small. In the present case, the probability for the earth to move away from the sun is so small that to write it down would require 10 to the power 10^{200} zeros to the right of the decimal point. The smallness of such a number staggers the imagination, and no computer could store it in decimal form. For all practical purposes, it is an event that will never take place.

As we move toward smaller objects, the probability of a tunnel effect increases. The probability for a car in a parking lot to move from one parking spot to another by a tunnel effect is as ridiculously small as that of the earth escaping from the sun's pull, but it has fewer zeros already. When my car breaks down, I know better than to blame it on quantum mechanics, the probability is still much too small. I rather look for a deterministic cause that a good mechanic will soon identify. However, as we approach the atomic scale the odds increase and quantum nondeterminism eventually overtakes classical determinism. In short, it is all a matter of scale. There is a continuous and quantitative transition of probabilities, from extremely small ones to others that first become non-negligible and later prevail.

Another feature of these theoretically possible but highly unlikely effects, of these "quantum fluctuations" that violate determinism, is that they cannot be reproduced. No quantum fluctuation observable on a human scale has probably occurred since the creation of the earth, but let us imagine one of them taking place and being witnessed by several people: they see a rock suddenly appear in a different place. They have actually seen it, but they would never be able to convince anyone else; never irrefutably show that the phenomenon may repeat itself. All they could say is "I swear it, the rock was over there, on my left, and all of a sudden it appeared on my right." Too much gin or whisky, some will say, a mild bout of madness, will think others, and the witnesses themselves will end up believing that they have been subject to a hallucination.

Thus, determinism is not absolute. We have also said that it ceases to be universal, in a sense that we shall now make precise. We have already seen that not all large physical systems are necessarily deterministic—take chaotic systems, for instance. The connections between determinism and chance in the case of chaotic systems are presently well known, and they form a vast domain of study that is beyond the scope of this book. We shall simply mention that quantum and classical mechanics both recognize the importance of classic chaotic phenomena, which mark the limits of an accurate correspondence between the classical and the quantum description of the world.

There is yet another condition to be met before we can trust determinism and common sense. It regards a system's initial state.

It is important to be able to specify this state as a purely classical property on which classical dynamics can be based. Now, there are cases, not so infrequent by the way, where this condition is not fulfilled. Here is an example. Imagine a Geiger counter isolated in a vacuum. This is a large-scale system that can be exactly described by classical physics. Determinism is particularly simple in this case, because it predicts that nothing happens. If we now imagine a radioactive nucleus inside the counter, the classical description of the counter no longer wholly captures the system's initial state (where system = counter + nucleus) and we must explicitly take into consideration the nucleus wave function. Since the fundamental laws of physics are quantum, the deterministic nature of the new system no longer follows. The entire wave function of the counter + nucleus system evolves according to Schrödinger's equation, and the fact that the counter is itself a metastable object, sensitive to small electrical effects, makes it impossible to establish in this case that the behavior will be deterministic.

In other words, the methods used to demonstrate determinism equally show that some exceptional cases exist in which determinism does not apply. The most frequent ones arise when measurements are taken on a microscopic object, as in the above example. This case is central to the interpretation of quantum mechanics, and for this reason we shall discuss it in the next chapter.

Hence, classical physics and common sense allow us to properly understand the large-scale world on one condition: we must not consider systems containing an instrument in the process of measuring some quantum object, or other even subtler devices. Put another way, we must restrict ourselves only to the situations known to human beings before the discovery of radioactivity, at the end of the nineteenth century.

A First Philosophical Survey

There is an essential result that we have repeatedly stressed: common sense conforms to the quantum nature of the laws governing the material world, at least in normal conditions and for objects on our human scale (and often even well below it), except in some extremely rare circumstances. Naturally, common sense cannot by itself determine its own limits of applicability, and for this reason the discovery of quantum mechanics was profoundly disturbing. We can only hope that this is only a temporary situation.

It is difficult, however, to fully appreciate all the philosophical consequences of such a result. Indeed, to imagine that common sense is merely the result of the laws of nature, and that these have their own logical structure, is a complete reversal of our usual patterns of thought. It is also difficult to get used to such a change in perspective, and its consequences are not always easy to apprehend. We can nevertheless draw a few simple lessons that have a direct bearing on the theory of knowledge.

To approach the knowledge of reality beginning with the laws discovered by science goes against traditional epistemology (it is, in fact, the exact opposite). We no longer try, as Bohr did, to use classical physics as our unique reference, as the only domain where logic can be applied and of which we can legimately speak. On the contrary, it is the quantum world that has its own rules of description and reasoning from which those of the classical world emanate.

We call into question the method followed by many people and preeminently John Bell, who sought to understand quantum physics through common sense, even if that required promoting certain of its aspects to the rank of philosophical principles (variously called "locality," "separability," "causality," and so forth). It is a completely opposite approach that turns out to be fruitful, one that has its foundations on the rock-solid principles of physics, painfully conquered by generations of researchers. From those principles we deduce the right form, the proper degree of approximation, and the domain of application of common sense. The latter then reappears purified and strengthened, no longer taken for granted without questioning and for this reason always mysterious. Restricted to its own sphere of application, common sense becomes a valid form of the laws of reality.

The above result also challenges the rules of philosophical inquiry. For it suggests that we no longer need to base it on the unbridled generalizations of our immediate experience denounced by Bacon, now that the patient efforts of research have paid off in the form of deeper principles, recognized by nature as its own, and close to the heart and the essence of things. Common sense, thus reappraised and with its scope circumscribed, no longer applies to the universe at large, and in particular ceases to be valid for the infinitely small. It would be vain for it to pretend to impose on the atomic scale philosophical "principles" that are simply the inordinate worship and the unjustified hypostatizing of our thought habits and language tics.

From the Measurable to the Unmeasurable

THE RECONCILIATION between common sense and quantum mechanics does not exhaust the lessons we may draw from the latter regarding the theory of knowledge. We have seen how quantum mechanics shut out common sense from the phenomena of the atomic world and this will lead to other revelations. Another major problem looms in the background: the relation between formalism and reality, between theory and nature, which will appear distinctly in the end.

To deal with the above questions we propose a method that rests solely on the principles of quantum physics, in particular on the logical ones. We shall proceed in a purely deductive fashion to guarantee the consistency of the enterprise, but that will not prevent us from discovering certain points of view that physicists and philosophers have missed until now.

THE POIGNANT PROBLEM OF INTERFERENCES

We have already come across an example of a quantum measurement in the previous chapter. It involved the Geiger counter that detected whether a radioactive nucleus had emitted an electron or not. By solving the Schrödinger's equation of the complex physical system formed by the counter and the radioactive nucleus, and assuming an intact nucleus at the start, we can establish the form of the wave function after, say, ten minutes. This function appears as the sum of two terms. The first one represents a nucleus still intact, with the counter still reading 0, while the second term represents a decayed nucleus and a counter displaying the digit 1 to indicate that radioactive decay has been detected.

We know that a wave function that is the sum of two terms allows, in principle, quantum interferences to take place between

196

the two states these terms represent. What happens in the present case? It is very difficult to imagine interferences between two different states of the same counter whose screen would display two different digits. Our imagination refuses to function, because *reality never confronted us with such a situation*. What's more, that kind of disagreement between theory and experience suggests that either the problem is not what it appears to be or the theory itself is unreliable. The second alternative takes us even farther: if interferences did exist, what would they look like? Two superposed photographs? Or perhaps two contradictory and overlapping visions like those provoked by a high fever? Since no such flickering of Reality exists, it is imperative that we should get to the bottom of it.

A lot has been written about this problem, which is often presented in a particularly striking form introduced by Schrödinger himself. It is worth repeating, even if it is well known and we have already mentioned it. A cat is trapped inside a box containing a diabolical device: a radioactive source whose decay will release a powerful poison. In its most straightforward form, the theory predicts that after some interval of time the cat's wave function will be the superposition of two others, one representing the possibility of an inactive source and a living cat, and the other representing the cat killed following the radioactive decay of the source. Many questions then spring up: Can we say that "the cat is dead" and "the cat is alive" are two separate events without interferences, without "flickering"? Can we be sure, without the shadow of a doubt, that only one of the two events has actually taken place, even if we cannot know which one without opening the box?

We can give another example, perhaps more explicit still, to illustrate the nature of the underlying difficulties. Imagine that a man called Pepin lived in Charlemagne's time. In one of the walls of his house there was a terrible radioactive nucleus. To keep the example simple, assume that only two events could have taken place. In the first case, the nucleus decayed when Pepin was three years old, causing his death; in the second, the nucleus was still intact when Pepin died, advanced in years, after having had children. Among the descendants of Pepin there were Napoleon Bonaparte and Professor Babillard, nowadays a specialist in quantum mechanics. The scientist studies the traces of the famous nucleus

and discovers interferences. What should he conclude? The interferences reveal the survival of the component of the wave function
corresponding to the situation where Pepin died when he was
three. Thus, still today there is a nonzero probability for Pepin to
have died while he was a child. The professor then begins his next
lecture as follows: "I have established that, in the present state of
the world, there is a nonzero probability for Pepin's death to have
occurred when he was a toddler. We must therefore reluctantly
recognize that Napoleon might never have existed and that I myself
do not exist either, for both events have a nonzero probability."

We can see where the difficulty is, if Babillard is right: no fact
could ever be conclusively confirmed. The very notion of fact, the
basis of every science, would be incompatible with the theory. Babillard's absurd statement is only slightly more exaggerated than
the assertions of those who would like to see in quantum physics
the grounds for a universal skepticism or for the wildest dreams.
Some talk of parallel universes, and consider the world in which
Julius Caesar is the father of Cleopatra to be as true as our own.
Others assume that only through human consciousness can the
sum of wave functions be broken apart. Still others go even further
and reverse the process: if our consciousness separates the possible
realities, then mind can act upon matter, and so parapsychology is
theoretically demonstrated. There are also those for whom science
is an imprecise, vague corpus where everything is possible, where
water has a memory that only wine can wipe out. Others, more
cautious, barricade themselves behind positions that they deem
more sensible: physics is only a convention among humans that
does not get through to reality; the wave function is nothing but
the expression of what I happen to know. Is it necessary to go on
and mention those who have founded on such gibberish not just
philosophies, but a psychology, and even theologies where God
contemplates all those multiple universes of his irresolute creation?

Bohr always attempted to preserve at all costs the objective
character of the science he had helped to found, and we shall see
that he was right in doing so. As for the rest, it is nonsense, twad-
dle, balderdash, and idle fancies (I also have some stronger words
in store). Wisdom would have consisted in saying, as an honest
Feynman or a doubtful Einstein did, "There is something we do
not understand." But then, you might wonder, how are we to
understand?

THE DECOHERENCE EFFECT

Much time was needed before an answer to the problem of macroscopic interferences could be found. There had been some hunches in early considerations by Heisenberg, but it was not until the 1970s that the correct answer was conjectured by Hans Dieter Zeh. His conjecture was tested on simple models, and orders of magnitude were obtained in the 1970s and 1980s by K. Hepp and E. H. Lieb, W. Zurek, A. O. Caldeira and A. J. Leggett, E. Joos, and H. D. Zeh. There could be no doubt: One of the quickest and most efficient effects in physics was at work in the vanishing of macroscopic interferences. More recently, a general theory with no reliance on special models was found by the present author. Still more convincing has been the experimental observation of the effects in 1996, in Paris, by a group headed by J. M. Raimond and S. Haroche.

The effect cannot be explained in simple words or by means of suggestive analogies. Otherwise, not so much time would have been needed for guessing the correct answer. The topic is, however, so important that we shall on this occasion overcome our reluctance to employ mathematical arguments in this book.

The story goes back to von Neumann, who noticed in the early thirties that the state of a measuring device appears very puzzling after a measurement has taken place, a remark that was later made popular with the famous example of Schrödinger's cat. Its basic assumption is that the measuring device and the measured system (an atom, for instance) both obey the laws of quantum mechanics.

Consider a device measuring a spin component of an atom. If the spin is 1/2, there can only be two possible values for the observable representing the spin component along some definite direction in space, the z-direction, say. This is the measured observable, and its two possible values are +1/2 and −1/2. Notice that both statements "the value of the z-component of spin is +1/2 (or −1/2)" are perfectly well defined by projectors in the framework of histories.

Von Neumann proposed a model for the measurement in which the measuring device is simply a needle pointing to a ruler. This pointer marks the position 0 before the measurement; more precisely, its wave function $\Psi_0(x)$ is very narrow and peaked for values of x very close to 0. When an atom enters the instrument with a z-component of spin equal to $+1/2$ (or $-1/2$), the pointer moves and its new position indicates the result of the measurement by pointing to $+1$ (-1) for a spin component of $+1/2$ ($-1/2$). More precisely, the wave function $\Psi_+(x)$ (or $\Psi_-(x)$) describing the pointer's position is very narrow and peaked for values of x very close to $+1$ (-1).

So far so good, and we might say that von Neumann's model gives a satisfactory description of a real measurement. But we are in serious trouble if the atom arrives with a definite component of spin of $+1/2$ in the x-direction, with the apparatus still ready to measure spin components in the z-direction. We find that the pointer's wave function at the end of the measurement is $\Psi(x) = (1/\sqrt{2})(\Psi_+(x) + \Psi_-(x))$. It contains a term $\Psi_+(x)$ indicating a pointer in position $+1$ and another, $\Psi_-(x)$, indicating -1.[1] Both properties are simultaneously present in the total wave function, expressing two different things, $+1$ and -1—a live cat and a dead cat. This simple formula holds the most difficult puzzle, the best hidden secret of quantum mechanics, and there can be no doubt that it is absolutely true. Many experiments, with an atom or particle in place of a pointer, confirm this conclusion: The superposition of wave functions is not a disease but it can easily be caught during an interaction.

Heisenberg and Zeh pointed out an oversight in von Neumann's model: A real measuring device is not a concept that can be fully described by one variable x. There are typically a few BBB particles in a piece of laboratory equipment (a BBB being defined for our purpose as equal to 10^{27}, or one billion times one billion times one billion). Therefore, the correct wave function should not be written $\Psi(x)$ but rather $\Psi(x, y)$, where y stands for some BBB variables or so. This wave function is the sum of two other functions $\Psi_+(x, y)$ and $\Psi_-(x, y)$, the analogues of $\Psi_+(x)$ and $\Psi_-(x)$, respectively.

The y variables stand for the microscopic features of the measuring device, including all the nuclei and electrons in it, and also often outside it: for instance, air molucules in the surrounding atmosphere as well as photons if, as usual, there is some light in the laboratory. By convention, the name "environment" has been given to the formal system described by the myriad of variables in y.

[1] For simplicity, we assume that after the measurement has taken place the atom has become lost among all those of the measuring device.

The intuitive idea behind the "decoherence effect" suppressing macroscopic interferences is as follows: A wave function such as $\Psi_+(x, y)$ is in fact a very complicated function of y and, moreover, very sensitive to the position of the pointer, that is, to the value of x. If the needle turns on an axis at the center of a circular dial, for instance, there is inevitably some friction, which can produce changes in the small world of atoms comparable to an earthquake on our scale. The two wave functions $\Psi_+(x, y)$ and $\Psi_-(x, y)$ are accordingly very different.

We may try to imagine one such function. Its sign changes in many places when only one among the BBB variables changes slightly, and in many more places when several variables change; the phase is practically random under the full impact of all BBB variables. Now think of two such functions that have suffered different fates. There is absolutely no chance that their signs, their phases, have anything in common for the same value of y, whence the name "decoherence," meaning that any phase coherence they might previously have had has been lost.

For macroscopic interferences to be seen for x, the observable parameter, it would be necessary that the dependence of the two wave functions on y be coherent. This is a technical point that can easily be proven but will be taken for granted.[2] The result at least is clear and simple: the erratic behavior of the wave functions in their description of the environment suppresses any possible manifestations of quantum interferences at a macroscopic, observable, level. If a cat is dead, its internal wave function will never regain the fine phase tuning of a living-cat wave function. Adding a dead-cat wave function and a live-cat one in a sum $\Psi_+(x, y) + \Psi_-(x, y)$ is like adding sea waves and the bubbling of a whale: they do not interfere, they ignore each other, they stay apart.

We cannot cast the theory of decoherence in this intuitive picture because we lack good theoretical tools for investigating the phase of a wave function with BBB variables, except in very simple models. Other techniques, inspired by the theory of irreversible processes and information theory, must be used. The main consequence of the decoherence effect is in any case most simply expressed in the language of logic: At any given time, Schrödinger's cat is either dead or alive, which is the most a basically

---

[2] Mathematically, $\Psi_+(x, y)$ is "orthogonal" to $\Psi_-(x', y)$, whatever the values of x and x'.

probabilistic theory can say. This is, however, in perfect agreement with common sense; there is no mystery here. We might add that this clear logical splitting may not hold during a short interval of time when the cat is dying (or when the pointer begins to move away from its initial position toward the still uncertain directions leading to +1 or −1; soon it will be going toward one or the other). One cannot say that the cat is both dead *and* alive, as was apparent in von Neumann's and Schrödinger's simplified versions. Decoherence has put an end to the old legend of Schrödinger's mythical cat.

THE WONDERS OF DECOHERENCE: PHYSICAL

The decoherence effect has many far-reaching consequences. Let us first look at it through a physicist's eyes. We have seen that it is strongly linked with friction or, more generally, dissipation effects through which energy is exchanged between the overall motion of a pointer, say, and the invisible thermal motion of its atoms. It should come as no surprise therefore if the general theory of decoherence provides also a theory of dissipation.

Accordingly, to the question, "What happens when there is no dissipation?," the answer is "No decoherence." Superconducting devices have been constructed showing an absence of decoherence in a remarkable way. Even if these SQUIDs (superconducting quantum interference devices) are macroscopic (they have the shape and the size of an old-fashioned hairpin), they exhibit a typically quantum behavior: tunnel effects. However, this kind of device is a laboratory curiosity, and there is little chance of finding one of them in nature. Much more common is a well-known macroscopic physical system impervious to decoherence: ordinary light. Radiation, when it involves many photons, is a macroscopic system in its own right. Photons interact in an extremely weak fashion, practically as if they did not interact at all, and there is accordingly no dissipation among them and no decoherence. We can thus expect to observe quantum interferences at a macroscopic level with their help. Do they actually occur? We know the answer, since these interferences were observed for the first time by Thomas Young.

Leaving aside exceptional cases, let us now consider the almost universal conditions under which dissipation can take place. Dissipation (or, equivalently, friction) tends to slow down, or damp, a motion. A pendulum, for instance, will not swing forever: it slows down. This effect is described quantitatively by a damping time, which is the time it takes for the pendulum's amplitude to be reduced by some arbitrary factor. The damping time is easily observable and the so-called damping coefficients are generally known.

Decoherence also has its own coefficient, its own decoherence time which is directly related to the damping coefficient according to the theory. This relation is such that the damping time is larger than the decoherence time by a factor involving the inverse square of Planck's constant. This is a tremendous ratio, which of course depends on other physical quantities such as the mass of the pendulum and the temperature. At any rate, Planck's constant is, in ordinary units fitting our own size, of the order of a BBB. Its square is thus a BBBBBB, or 10^{54}. Since the damping time is generally sizable, the decoherence time is correspondingly extremely small, so that decoherence turns out to be a tremendously efficient and rapid effect.

As an illustration, let us look at an extreme case where decoherence is particularly slow. This can happen at zero temperature in vacuum. Consider then the case of a pendulum that starts as the superpositon of two states, like the pointer we discussed earlier. The two positons are separated by a distance of one micron. We assume for definiteness that the pendulum's mass is 1 gram, its period is 1 second, and the damping time 1 minute. The efficiency of decoherence can be expressed by the time it takes for macroscopic quantum interferences to be reduced by a factor of 2. The answer is 10^{-16} second: the effect is undoubtedly very efficient! This conclusion is further confirmed by realizing that in this case the effect is particularly slow (if the word is appropriate). At nonzero temperatures in an external environment macroscopic interferences disappear much more rapidly. For instance, it is enough to have a few air molecules collide with the pendulum for interferences to begin vanishing. Decoherence is, moreover, very active: it begins gnawing at quantum interferences right away, without giving them enough time to develop.

For a long time one was left with the perplexing impression, on the evidence of explicit models, that decoherence is so expeditious that it would be impossible to see it. Or, more precisely, that our experimental tools could never catch the effect in action; they are too slow, they arrive too late, after interference has taken place and no trace of it is left. Physicists do not like an effect to be so elusive. They want to actually observe it before being completely convinced. How can this be done? Obviously by using an object that is on the verge of being macroscopic but is nevertheless still microscopic—known as a mesoscopic object.

Such a system, a very clean one, was used by Raimond, Haroche, and their team. They brought a rubidium atom into a state with very high quantum numbers, where the electron is very far from the nucleus. After traversing a suitable device, the atom comes out in a superposition of two states, ready to show interferences. The measuring apparatus may be roughly described as a "radiation pointer." The pointer consists of a few photons (from 1 to 10) in a cavity, whose walls form the pointer's environment. To make a long story short, let us just say that quantum interferences can be observed, and one can see them decrease with time according to the theory. The duration of the decoherence varies with the number of photons in the cavity and with the value of some other parameters that can be adjusted. In conclusion, there is no doubt that decoherence is a real physical effect that destroys quantum interferences at the macroscopic level. Moreover, we can understand it as a direct consequence of the basic principles.

THE WONDERS OF DECOHERENCE: LOGICAL

Decoherence also has some basic consequences for epistemology, and even for the philosophy of knowledge. This aspect of the story began with a question due to Wojciech Zurek, one of the most active investigators in the theory of decoherence. The very sketchy mathematical description of decoherence we presented earlier can be repeated for any wave function $\Psi(x, y)$, without splitting it into two preferred states Ψ_+ and Ψ_- (representing a live cat and a dead cat, or the states of the needle pointing toward -1 or $+1$ on a dial). Why not look directly at the full wave function $\Psi(x, y)$? Decoherence would then predict that two functions of the unknown BBB environment coordinates y such as $\Psi(x, y)$ and $\Psi(x', y)$, for fixed values of the macroscopic coordinates x and x', will lose phase coherence sooner or later. This tendency cannot go as far as letting x be equal to x', since $\Psi(x, y)$ has obviously a full phase coherence with itself. Wojciech's point was that this kind of "diagonalization" (that is, the loss of coherence except for x almost equal to x') breaks a rule of invariance that was most emphasized by Dirac in the late twenties.

The Lagrange–Hamilton version of classical dynamics is invariant under a large group of canonical transformations involving both position and momentum variables. Similarly—and this was Dirac's point—quantum mechanics is invariant under its own large group of canonical transformations, which is very useful in many theoretical investigations. The kind of decoherent diagonalization we were discussing cannot, however, be invariant under the full group of canonical transformations (in the x variables and the corresponding momenta). It must select some special variables (a specific "basis" in a more technical language), x being, for instance, a position coordinate and not a momentum, or anything else.

We cannot go into the mathematical details needed to explain why our old friend, the principle of inertia, is ultimately responsible for a particular choice of variables in decoherence. We shall only state the result: Decoherence leads directly to Newton's conception of classical mechanics. Who cares? you may be thinking, Newton or Hamilton are the same to me. Please bear with me, for it is really wonderful.

It should first be said that there was something amiss with the recovery of common sense and classical physics we presented earlier. We had been able to recapture classical mechanics, but in a rather abstract form, along the lines of Lagrange and Hamilton's mathematical formalism. True enough, we had recovered common sense and determinism, but I did not tell you, dear reader, how abstract this view of our world remained. I simply said, mischievously, that the angel was satisfied with it. An angel, perhaps, but what about you and me? Are we used to walking in a so-called symplectic phase space or merely in the street?

The results on decoherence cover the remaining stretch of the road leading from the abstractness of quantum principles to the cosy comfort of common sense. One can include, for instance, the dissipation effects in a classical description of dynamics. The last step back to the sources of physics, from Lagrange to Newton, is, however, much more significant. Newton thought of a macroscopic mechanical object, either solid or fluid, as made of small parts located in ordinary space, not of course in a mathematical space with as many coordinates as there are degrees of freedom. These small pieces of matter are out there, though containing many atoms. I can put my finger on them and tell you, "Look at this." Decoherence, when the principle of inertia is taken into account, tells us that this vision of the world in ordinary three-dimensional space, or rather the validity of this vision, is truly a consequence of quantum mechanics. In our own world we can be simple again, Greek again, with a clear mind. We can see. The visual representation of our macroscopic world is in perfect agreement with the basic laws of physics that seemed to negate it for so long, and harmony is restored. I guess this calls for a celebration and might well be worth a glass of champagne.

When the angel was told of this result, which has been known in Heaven since BBB millenia or so, she was exhilarated. She could at last open her eye, which knew only the photonless light of that higher place, and see our world far below. "How beautiful is earth," she said, "when seen as human creatures can see it. How pale a wave function is when compared with the rose hidden in it. This way of looking at the world is the only real one. It is the way of love."

When there is a sudden moment of silence during a conversation among several people, the French say that an angel has gone by. We may perhaps have a short moment of silence to let our angel go. And now that we are among human beings, is that everything decoherence can tell us? Not at all. You may remember the crazy story of Professor Babillard, who discovered that he might not exist because of quantum mechanics. The story appeared crazy to anyone with common sense, but it turns out now that it is hopelessly crazy, against logic, if you prefer, because it can be expressed in terms of some Griffiths' histories that can be shown to be inconsistent. This inconsistency, namely, a failure to satisfy Griffiths' consistency conditions, can be proved from decoherence. But who cares about Babillard? The result is much more significant: any property that can be asserted as a consequence of decoherence will afterward remain valid forever; it cannot be invalidated by later events. This means that the concept of fact is perfectly valid in quantum mechanics. If one adopts Bohr's definition of a phenomenon as a conceivable fact, then all phenomena can be considered as classical properties resulting from decoherence.

When we remember that common sense directs not only most of our thinking but also our actions, this possibility of relying on facts is of course essential. Decoherence fully saves the appearances of ordinary reality.

LAST WONDERS: THE DIRECTION OF TIME

The last wonder offered by decoherence has to do with the direction of time. An old question in physics concerned the preferred direction of time in the world we see around us: A pendulum slows down and its oscillations do not increase spontaneously; divers fall down and do not fly out of the water; we realize immediately that a film is being run backward. Time has a direction in a macroscopic world, but there is no such privileged direction among particles. The basic laws of physics, including quantum mechanics, are the same when the direction of time is reversed. How can these two conflicting facts be reconciled?

We have already stressed the relation between decoherence and dissipation. Decoherence operates in a specific direction of time, which is linked with the events it relates. It is impossible to go, for instance, from the state of a dead and live cat, as found formally in Schrödinger's experiment, back to the live cat we had at the beginning. Decoherence is logically inconsistent with such a reversal of events. Why? Because we would not only have to simply prepare a cat that is dead and another that is alive, which is easy, but we would also have to prepare exactly their internal wave functions, down to each of their BBB variables, for their state to go back to the one Schrödinger assumed in his box. This is impossible.

Is it absolutely impossible? Not quite. If the "cat" is made up of two or three atoms, the reversal can be performed in some cases, and the direction of time does not matter. But with BBB atoms, it cannot be done. Or, more precisely, the necessary experimental device to perform the task of preparation would be bigger than the whole universe. So big that it would not work, due to special relativity: actions in the apparatus would take a practically infinite time. However, some say that this is a question of principle. Why not consider a device composed of an imaginary sort of matter so as to be small enough, though still with the necessary number of "degrees of freedom"? It would be small indeed, but so heavy as to collapse immediately into a black hole. In a nutshell: the laws of physics we know forbid a change in the direction of time for a sufficiently big cat.

Decoherence selects a specific direction of time for the events it can link in a consistent way. Because of the close relationship between decoherence and dissipation, this direction of time is the same as in thermodynamics. And finally, because decoherence is by far the most efficient mechanism for ensuring the validity of quantum logic (Griffiths' consistency conditions), there is also a specific direction of time in our logical account of the world, in the common sense following from it, and this direction is also the same as in thermodynamics.

The word "wonder" is certainly appropriate when we realize the way in which quantum mechanics deciphers secrets that went almost unnoticed: that of the validity of classical physics and the value of common sense. How far we are from Hume's surrender and Kant's restrictions, and how clearly this should point the way toward new vistas in philosophy. A world of the intellect where humankind fully appreciates the consequences of the most efficient effect in physics can never be the same as the old one.

MEASUREMENT THEORY

We have said that measurement theory proceeds in a deductive fashion from first principles. Some of the results we have already mentioned play an essential role in this deduction. Thus, the experimental data displayed by a measuring device may be described in a purely classical manner—we saw why this is possible when we discussed the retrieval of common sense. It is also necessary that the data be free from interferences, this being a consequence of decoherence. The latter also plays a major part in the fact that the data belong to consistent histories (in Griffiths' sense).

In order to properly understand what a measurement is, it would be helpful first to make a distinction between two notions that are frequently confused: an experiment's (concrete) data and its (meaningful) result. The data are for us a macroscopic, classical fact: thus, when we see the numeral 1 on the Geiger counter's screen, this is the *datum*. The *result* is something different, for it is a strictly quantum property, almost invariably pertaining only to the microscopic world, meaning that a radioactive nucleus disinte-grated, for example, or providing a component of a particle's spin. The datum is a classical property concerning only the instrument; it is the expression of a fact. The result concerns a property of the quantum world. The datum is an essential intermediary for reaching a result.

A rigorous theory must begin by specifying the attributes that make a given experimental device into a measuring instrument. We shall omit them, though, and stay away from technical details. What matters is that from those criteria we can establish the key to our conclusion: *the datum and the result are logically equivalent.* This equivalence may excuse those who had never made any dis-tinction between the two, even if this theorem takes advantage of the full force of the theory's logical and dynamical formalisms. It is also an example of quantum logic's amazing power and of its po-tential for clarity. Let me emphasize that this theorem rests only on the following hypotheses: we are dealing with a measuring instru-ment that we assume to be perfect (imperfections may be ac-counted for later); the instrument is subject to the decoherence ef-fect; the rest of the hypotheses are the principles of the theory.

Another important result regards probabilities. We may put it roughly as follows: by repeating the same measurement a large number of times we obtain data amenable to statistical compila-tion, the result of which is necessarily in agreement with the the-ory's elementary probabilities as they were postulated from the start. Recall that in our theoretical construction those probabilities appeared simply as a logical, or linguistic, tool. It is only at this stage that they finally acquire the empirical significance they were lacking, and that chance enters the theoretical framework. We have thus achieved the point where the theory may finally be com-pared with experience, and the road leading from formalism to concrete reality is at last complete.

WAVE FUNCTION REDUCTION REVISITED

One of the most important rules promulgated by Bohr concerns two successive measurements. In its weakest form, this rule states that the probabilities of the results of the second measurement may be computed "as if" the result of the first one determined the wave function at the exit from the measuring instrument. The exact form of this wave function is irrelevant for us (for the theoretician, it is an "eigenfunction" of the first measured observable). The question is to find out whether the above "as if" conceals a practical recipe or a physical reality. The rule itself has at any rate been widely confirmed through numerous experiments, so that its validity cannot be questioned.

We have seen that Bohr did not consider the reduction rule to be a mere empirical prescription, but as one of the most fundamental laws of quantum mechanics—an authentic law of nature. He even considered it to be unlike the other laws, for it alone allows the theory to be applied, and hence to be verified. He also thought that it was impossible to subject it to empirical verification, for the rule was a prerequisite for any prediction. He even placed it on a more fundamental level than Schrödinger's dynamics, given the fact that the latter ceased to apply when a measurement took place.

The answer provided by the new approach is much more prosaic. In it, wave function reduction does not appear as a truly physical effect, and it is not necessary to even mention it in order to develop a complete measurement theory. Indeed, within the framework of Griffiths' histories, and taking into account the complete histories of the measuring devices as well as those of the measured objects, nowhere do we find something resembling a wave function reduction, and everything remains in perfect agreement with Schrödinger's equation. We observe only a purely mathematical result: the probability of a history involving two successive measures *may* be written in a form similar to the reduction rule, and which, moreover, generalizes the latter when it is not well defined.

Thus, wave function reduction would be no more than a convenient but nonessential recipe, a circumventing formula that allows us to bypass a logical calculation. The rule appears when we disregard the detailed history of the first measuring device to consider only the datum that it has produced; we then follow the history of the measured atom as it enters the second instrument and obtain in this way the result predicted by the rule.

There is a striking similarity between this result and other more familiar forms of logical simplification. In our discussion of logic and mathematics, we have seen that it is permissible to forget the steps in the proof of a theorem and recall only its conclusion, which we may then use as the starting point for new proofs (this is what we called *modus ponens*). Wave function reduction is in a sense a kind of *modus ponens*, another logical shortcut where entire portions of the measuring instruments' histories are deleted. The only true physical effect that conditions the result is the decoherence that actually takes place in the measuring device—and not in the object being measured, as was long believed to be the case.

THE CHASM

From what we have just said it might appear that quantum physics' first principles generate their own interpretation, and that they lead naturally, without any additions, to an image of the ordinary world that is in perfect agreement with its most familiar features. May we finally relax and say that everything is simple? Unfortunately (or fortunately?), no. For we have to deal with a monumental question, that of a reality seemingly wishing to tear off the wrapper of thought with which we have covered it. I call this immoderation the chasm, for it is the wide-open mouth of the Abyss; not the mouth that speaks to Hugo, but the one that growls.

Chasm, where do you come from? Einstein shivered at your sight and rejected you: No, "God doesn't play dice!" We shall approach you, the formidable one, but cautiously.

Let us speak as physicists and return to the wave function reduction. When we said that the reduction rule could be deleted from the list of principles of the theory we ignored the fact that the rule hid a difficulty that is always present: every measuring experiment results in a single datum, in a tangible, unquestionable fact. Now, against that, what do we have to propose? A theory woven from probabilities, a game of possibilities. Nothing in our theory offers a mechanism, a cause, from which the virgin present, the immutable and pure uniqueness of Reality would result.

The greatest questions dazzle, and numbers of physicists prefer here to cover their eyes. They hide away in the reassuring den of the theory, from which they refuse to come out. The theory, some of them argue, contains all possible worlds; we shall therefore conceive an immense wave function, born with the universe and evolving ever since in conformity with the quantum laws. Each time an alternative opens up, the universe's wave function branches out to conform to all possible outcomes. It takes a trifle, almost nothing; a nucleus, in some obscure planet, disintegrates (or not) and leaves a trace (or not) on some inaccessible rock and the majestic wave function of the entire universe bifurcates. Same result if a physicist measures a quantum effect in the laboratory. A pebble is pushed right instead of left by the thrust of a mountain torrent and the function forks again. Some of these events may be grandiose; perhaps a little more matter or radiation here instead of there when the universe was very young might result, in the remote future, in two galaxies of different shapes; thousands of stars would be affected. But most events are insignificant, of minimal consequences.

Sure, but that is to be expected in a world where chance plays a part. The theory is perfect because it incorporates chance; it conceives only the possible. We shall mention, without endorsing it, a strange idea proposed by Everett in 1956. Everything the universe's wave function contains since the beginning of time is not, according to Everett, a cemetery of ancient possibilities, never realized, of which the unique survivor is the world we see today. The wave function conforms to as many parallel realities as there are possibilities, each following its own separate course. Reality is not unique.

Insane idea, you might think, and I would agree. Everett's conjecture appears to be the wild dream of a mind intoxicated by theory rather than the product of a sensible reflection. And yet, can I completely refute it? Certainly not! From what we know about decoherence, an entity or being in one branch of this multiple reality may never reach another branch; no experiment can establish that other branches exist as well, or that its branch is the only one. The parallel universes, in their uncountable multiplicity, completely ignore each other.

That settles it, an empiricist would say; it is therefore pointless to discuss the topic any further. Science only studies verifiable facts, and those theories cannot be verified. Hence, they have nothing to do with science. Let philosophy deal with them, if it so pleases.

It is precisely the point I wanted to make. Since Everett's theory exists and cannot be refuted, even in principle, the question of the uniqueness of Reality does not belong to the domain of science, of what can be verified, but to that of philosophy or metaphysics.

That being said, why not philosophize a little? Let us examine the position opposite to Everett's, which must be called metaphysical despite all its appeal: Reality is unique. "Things are what they are; profound, profound is that. Before he who prostrates himself, we shall prostrate ourselves" (Lubicz-Milosz).

Profound indeed is that, but let us put it in milder terms, in the form of a law of physics such as "Reality is unique. It evolves in the course of time in such a way that different events that occur in identical circumstances have frequencies of occurrence in agreement with their theoretical probabilities."

Formulated in those terms, the idea is not entirely new, Niels Bohr having anticipated it under a similar form. Remember the special role he attributed to the wave function reduction rule. He distinguished it from the other laws of physics; it was for him the foundation of the very possibility of comparing theory with experience, and so it eluded any experimental verification. Surely, we have later seen that the practical wave reduction rule is not the expression of a physical effect but a simple logical convenience. However, the rule had for Bohr two quite different meanings: first as a practical rule, to compute the probabilities of the outcomes of two successive quantum measurements—which later became a simple theorem. But the rule also accounted for the occurrence of a unique event among all possible outcomes of a measurement, and it is precisely for this reason that it was different from the other rules. Interpretation has much progressed since Bohr's time, and it is now clear that the rule we have stated above captures the essence of his ideas, even if its form is very different.

And now, here we are facing the chasm. What does this very special, metaphysical rule say if not that theory misses what is perhaps the essence of reality? Every characteristic of reality has reappeared in its reconstruction by our theoretical model; every feature except one: the uniqueness of facts. Theory and reality agree on every aspect but for that single hiatus. Their disagreement is, however, absolute (and I do not employ this word lightly), for this conflict occurs at the most intimate level and each opposes the very essence of the other. Theory, being purely mathematical, can embrace only the possible, and its probabilistic character is inescapable. Reality, on the other hand, is above all unique, for it is what is completely defined when we point our finger and say, "That."

We seem to have reached a limit, some fundamental barrier that cannot be crossed; a warning, gravely telling us that the forms that mathematics and its Logos can express do not fully fit reality. What can we say other than that we have attained the bounds of the "Cartesian program," denounced only by Heidegger and until now a complete success?

During more than half a century, countless philosophers and physicists have reproached quantum physics for not *explaining* the existence of a unique state of events. It is true that quantum theory does not offer any mechanism or suggestion in that respect. This is, they say, the indelible sign of a flaw in the theory, implying that a better theory should replace it in the future. In my opinion, this attitude originates in an idolatry of theoretical explanations. Those critics wish at all costs to see the universe conform to a mathematical law, down to the minutest details, and they certainly have reason to be frustrated. For a long time everything seemed to be going their way, but listen to the chasm growl. Come, you mortals, and look at Reality, at what is, at what is flowing in a river where nothing is ever in the same place twice, at what is endlessly creating and changing; look at all that and now dare reduce it to a mere appendix in the Logos of your mathematics, from which time is barred and where stillness dwells forever!

I embrace, almost with prostration, the opposite thesis, the one proclaiming how marvelous, how wonderful it is to see the efforts of human beings to understand reality produce a theory fitting it so closely that they only disagree at the extreme confines. They must eventually diverge, though; otherwise Reality would lose its nature proper and identify itself with the timeless forms of the kingdom of signs, frozen in its own interpretation. No, science's inability to account for the uniqueness of facts is not a flaw of some provisional theory; it is, on the contrary, the glaring mark of an unprecedented triumph. Never before has humanity gone so far in the conquest of principles reaching into the heart and the essence of things, *but that are not the things themselves.*

214

Addendum

Some progress has been made on the problem of uniqueness since the publication of the French edition of this book.[3] It is mainly a matter of logic: it was shown that the present interpretation of quantum mechanics is perfectly *compatible* with the uniqueness of Reality; that is, this uniqueness is not predicted by first principles but it does not contradict them either. There is therefore no problem per se, nothing to be solved by a new theory, but only the wonder of theory and Reality fully agreeing even if their essence is ultimately different.

Another remark may be added. Some physicists persist in seeing as a significant problem what they call "objectification" (a rather ugly name, by the way). It may be stated as a question: How is a unique datum produced when an atom or a particle interacts with some measuring device? As a matter of fact, there is no language endowed with logic in which this question makes sense. Just like the famous question, "Through which hole did the particle go?", this illusory problem is only a treachery of our classical mind: the mirage of common sense creating visions where there is really nothing sensible.

[3] See the author's *Understanding Quantum Mechanics* (Princeton, N.J.: Princeton University Press, 1999).

On Realism

THE RELATION between quantum mechanics and realism has always been a matter of controversy. Much has been written about it by great thinkers. Their shadows over our head should invite us to be cautious, but the topic cannot be ignored in any philosophical analysis.

A BRIEF HISTORY OF REALISM

The basic question is simple: Is scientific knowledge a knowledge of reality? Realism stands for an affirmative answer. When science explains that a table is made of atoms, how sap ascends in a tree, or the workings of our heart, it is saying, or appears to be saying, what these objects really are. Bernard d'Espagnat carefully analyzed the doctrine of realism, which he defined in more general terms as a belief: to consider as obvious that "something" exists, a reality whose nature does not depend upon our cognition faculties or our actions when we are observing or measuring. The doctrine of *physical realism*, which is the one giving rise to controversy, adds a stronger assumption: through scientific investigation we can have access to a proper knowledge of this reality, at least in principle.

Many people believe that physical realism was typical of classical thought, before new questions were raised in the twentieth century. However, the history of science does not support this sweeping view but tells something rather different.

Science was still very young when some scientists began to call its meaning into question. Another doctrine was proposed that rejected realism, either partially, and just as an afterthought, or even completely. It adopted many different shades in the course of time but basically it states that science provides a description of reality in which the appearances of phenomena are preserved. Phenomena are understood as something we can see or touch and, more generally, reach through perception or experiments. The word "appearances" also refers to our perceptions, which are taken for granted (scientists rarely consider spells of illusion worth discussing). According to this wide-ranging doctrine, science provides a representation of phenomena but does not attain a knowledge of reality itself. An extreme position considers that science produces "only a representation," while more moderate ones express in various ways the shades between exact knowledge and efficient representations, depending on the circumstances.

The conflict between realism and the doctrine of representation was analyzed by Pierre Duhem (1861–1916), a great historian of science who also made significant contributions to thermodynamics and physical chemistry. His works are particularly interesting because written during the period when physics was just turning away from classicism. He knew Planck's and Einstein's works but was not much influenced by them. He was, however, well aware of a change in the spirit of physics as a result of Maxwell's contributions, and he could foresee the dominance of formalism. Since the main developments in quantum mechanics occurred only after his death, Duhem is an excellent witness of classical physics meditating on its own nature.

According to Duhem, there is a close relationship between realism and explanation, already found in Aristotle. In the ancient world, "physicists" were defined as those trying to explain things as they are, as our eyes see them. A good example is provided by the atomists, who explained optical phenomena by means of atoms of light flowing from the sun, bouncing off illuminated bodies and entering our eyes. An explanation of these phenomena by the "physicists" then consists in forming in the mind a clear image of the things as they *are*, atoms being, for instance, very small bodies similar to grains of sand or specks of dust. The same image can then be recreated in another human mind using words.

Classical realism is more or less that: it assumes that everything real can be understood, seen by the eyes of our mind, and accessible to the power of reason. When Boscovitch (1711–1787) revived atomic theory, he offered an explanation of new phenomena, such as the effect of pressure and some chemical properties. He had no doubt that atoms are really as our imagination sees them. He is a true realist. When Descartes asserts that matter coincides with space, he "sees" it, and is sure that his imagination agrees with reality. As for Galileo, we must consider him a less extreme and

217

rather cautious realist. Descartes, Galileo, Boscovitch, anyone else? Duhem made a list of the scientists who were clearly realists and it practically stops there. In the view of all other authors, realism is mixed in various amounts with a belief in the features and existence of a representation.

The initial signs of this cautious position appeared very early, with the first theories in astronomy. Hipparchus made a remark we have already mentioned, namely, that two different theories may account for the planetary motions, using either epicycles or eccentrics. How can we choose one over the other? Perhaps both should be suspected.

This doubt persisted, from Posidonius (around 131–51 B.C.) to Saint Augustine and Simplicius in the first half of the sixth century. Saint Augustine expressed it very clearly. Speaking of the motion of Venus, he said, "Astronomers have tried to express this motion in various ways. But their assumptions are not necessarily true since the appearance one sees in heavenly bodies might perhaps be saved by some other form of motion yet unknown to man." The famous motto according to which science "saves appearances" occurs many times in the history of ideas and it was used by Duhem as a title for one of his books.

When Copernicus' theory was published, the same question was asked: Does the earth really revolve around the sun, or is this only another way of saving appearances? If so, it would be simpler than the previous ones, since it needs a smaller number of epicycles, each one of them of smaller size. When the Catholic Church realized what was at stake in Galileo's time, it advocated a purely representative conception of science over complete realism, according to Thomas of Aquinas' philosophical views. Galileo was condemned for refusing this concession: he was, after all, a realist.

Our present views on this matter are closer to Thomas of Aquinas' than to Galileo's famous "But it does turn." One of the basic tenets of our present theory of gravitation is that the laws of nature are the same irrespective of the reference system used, from which it turns out that the explicit equations for the laws are rather simple. However, such a simplicity criterion has no decisive objective meaning, since only the mathematical *form* of the law has a universal value, and this form embodies all the special representations one may wish to use to describe the phenomena. Realism will never be as simple as it used to be, and its "explanation" of the world must be in any case much less conventional.

Leaving aside these modern considerations, let us go back to classical physics and listen to what Newton had to say on this topic. In his *Principia*, he said that with the help of his force of gravity he could explain phenomena in the sky and the seas, but could not assign any cause to gravity. Though convinced that such a cause exists, he did not feign a hypothesis for it, since anything that cannot be drawn directly from the phenomena must be recognized as a hypothesis. By his abstention, Newton moves away from realism, leaving us with a representation of reality by principles with no further foundation. In the second version of his *Optics*, he says that principles are only a condensed summary of observed phenomena.

Newton does not exclude subsequent access to some realism, though; he only refrains from saying more, in the belief that a direct knowledge of the created world is reserved to the creator.

He is most careful not to follow Descartes, whom he thinks presumptuous for having said, "As for physics, I would consider knowing nothing of it if I only knew how things can be without a proof that they cannot be otherwise." To which Pascal had angrily answered, "One must say roughly: this is made by shape (figure) and motion, because it is true. But to say which and build up the machine is ridiculous; it is useless, never ascertained and laborious." This peremptory judgment was to remain a threat for all later attempts at realism.

Everything had therefore already been said in the seventeenth century. A reasonably complete discussion of classical realism would have to include more, but we shall be extremely sketchy and only mention Kant's cogent objection: all knowledge must go through the mold of our a priori synthetic judgements, the constraints of our mind, so to speak. *Noumena*, the things themselves, are inaccessible. This is not very different from Bohr's point of view, formulated much later.

Much could be said on the representative value of models in nineteenth-century physics. Take, for instance, Maxwell's extremely fruitful mechanical model for ether—even if a model of something that does not exist. It was to be replaced by principles whose summaries are only equations. We should not forget this

219

remarkable episode, which shows that realism can sometimes serve as a guide in science, but that strict rationality and logical consistency may well have the last word.

We should also remember thermodynamics, the science that tells us what can be said when we know practically nothing about a system. Our present statistical mechanics is much more realistic, although one should not forget that it has its roots in quantum mechanics. Mach's and Avenarius' speculations on the correspondence between our mental processes and physical phenomena also indicate the difficulties of realism. Modern cognition sciences have shown how nontrivial and subtle our perceptions are when decomposed in our sensory organs before their recomposition in our brain. Yesterday's explanatory "images" have now lost much of the obviousness they had.

Modern science has considerably increased the scope of realism. We know the structure of a crystal and of DNA molecules, and we think we know the internal structure of the sun. We have made many direct observations and have clear mental pictures, except for foundations that are only formal, as is the case in quantum mechanics. This leads to a distinction between two notions of reality: There is ordinary reality, things we can touch or see, often with the help of our instruments. There are also things we consider as real though we cannot have a picture of them: atoms, particles, quantized fields. We know only their laws, which are still in a sense a summary of experimental facts.

Bernard d'Espagnat took the bold step of considering that in such cases the concept of reality may apply directly to the laws of nature themselves. There is a "realism of accidents," for events occurring more or less by chance or fortuitous circumstances, valid for ordinary reality. However, its range does not extend to the quantum world, where the laws stand as exhaustive. Renouncing the knowledge of microscopic accidents casts a veil on their reality, of which no corner can be lifted.

Finally, we should not forget that in the history of ideas, realism was an important topic among philosophizing theologians. Saint Agustine was interested in that question only as a preliminary for theology, while some of our modern questions may remind us of other theologians such as Denys the Aeropagytus, Master Eckhart, Saint Bonaventure, and Nicola da Cusa. They all took for granted that God exists, but were so exacting and had such awe and respect that they thought him far beyond what words can reach. Inexpressible, unutterable, the Greek word for it is "apophatic," and this tradition is called apophatic theology. Perhaps realism is ultimately an apophatic epistemology. Is this a joke? Maybe.

Quantum Physics and Realism

Einstein was particularly worried by the lack of realism in quantum mechanics. Why this absence of realism? A mathematical argument, which we cannot develop in detail, would go essentially like this. Suppose we are interested in the position of a particle at various times. According to quantum mechanics, there is a wave function expressing the probability of finding the particle in a given place. The evolution of wave functions in time is described by Schrödinger's equation, and the wave function at time 1 depends linearly on the wave function at time 0. Suppose now that the particle is *really* in a specific place at time 0, although we do not know exactly where. We do not know the particle's velocity either but, if the particle is in some place at time 0, it will be (really) in some other place at time 1. According to Laplacian theory, a basic rule of probability is the theorem of composite probabilities which, in the present case, says that the probability for the particle to be somewhere at time 1 is a linear combination of the probabilities of its possible locations at time 0. The coefficients in this relation are the probabilities for the particle to go from one place to another between times 0 and 1.

Now comes the contradiction: the probability is given by the square of the wave function. It is then impossible, except in very special cases, to have simultaneous linear relations between both wave functions and probabilities. Therefore, it seems that the basic assumption, namely, that the particle is really in some place, should be wrong.

This negative result is of course intimately linked with the impossibility of assigning a trajectory to the particle, which is the main consequence of Heisenberg's uncertainty relations. We might add that Griffiths' consistency conditions for histories often select precisely the "special cases" mentioned above. We shall have more to say on this later.

Some physicists who were strongly motivated by realism tried to find a way out. David Bohm, for instance, said that every particle has a definite position and momentum but that its motion depends on a wave function. The previous argument then comes apart, because it would have to take care not only of the "real" events at times 0 and 1 but also of a "realistic" wave function. This direction of research is still active, although it has not yet answered questions such as these: Are photons real? Is the electromagnetic field real? Quantum mechanics needed only one year to go from quantum electrons to quantized radiation, but this nagging problem still remains unsolved more than thirty years after Bohm's initial attempts. And Nelson's stochastic quantum mechanics approach, where the relation between probabilities is changed, has not been any more successful.

Complementarity was at the basis of Bohr's argument against naive versions of realism. As we indicated within the framework of histories, complementarity shows that one may consider some properties of a logical system and deal with them in a logically consistent way, but that often there are completely different consistent histories incompatible with the first ones, introducing, for instance, a property of momentum rather than a property of position. Both are logically legitimate descriptions, but they exclude each other. Thus, one cannot speak of a real property.

Bohr swept aside most questions on reality, although he insisted on the objective character of quantum mechanics. "[Quantum mechanics]," he said once, "requires a final renunciation of the classical ideal of causality and a radical revision of our attitude toward the problem of physical reality." Elsewhere he said, "In our description of nature, the purpose is not to disclose the real essence of the phenomena but only to track down, so far as it is possible, relations between the manifold aspects of our experience." And as a final blow: "We must never forget that 'reality' is a human word just like 'wave' and 'consciousness.' Our task is to learn to use these words correctly—that is, unambiguously and consistently."

Up until now, the relations between the manifold aspects of our experience have all been found contained in the principles of quantum mechanics. We have also seen that the formal character of the laws brings out another facet of physical realism: its relation with the nature of mathematics. This aspect might open the way to some sort of grand realism where the whole field of science is concerned, and some of d'Espagnat's considerations go in that direction. Bohm's and Nelson's approaches, which search for an ontological status for accidents, may be called petty realism (no slight intended). The inclusion of mathematics within reality would be a much wider kind of realism, and one which we later intend to uphold.

ORDINARY REALITY

We have already defined ordinary reality as everything we can see or touch. It consists of the obvious things Wittgenstein's bricklayer can point to while saying "that" to his assistant, without any ambiguity.

Has ordinary reality a place in a world governed by quantum laws? The answer is definitely yes. The things we can touch or see, even with the most powerful microscopes, are macroscopic. We have already seen how common sense can deal with them when considered from the standpoint of quantum mechanics. Moreover, the properties of these things that we perceive are immune from the ambiguities associated with complementarity.

Expressed in more technical terms: although the basic laws are quantum mechanical, the properties and phenomena occurring in the macroscopic world can be stated classically, and it is legitimate to do so (this result has by now been completely established). When a phenomenon is observed, we call it a fact. Since we are not solipsists, we also admit that many facts exist everywhere though no observer can see them. Facts are said to be true.

An essential feature of our language is its capacity to deal with possibilities as well as with facts. Facts are actual and phenomena are possible, and statements about them entering in verbal propositions are either true or false. This notion of truth or falsehood is legitimate from a logical point of view because, in spite of complementarity, classical statements that are meaningful are unambiguous. Some of them can be proved to be true from the mere observation of a fact. For instance, I leave a book on a shelf, close the door, and make sure nobody enters the room, no hurricane blows through a window, and a few similar conditons are satisfied. I can then assert as true that the book is still on the shelf although no one can see it—this could even be proved as a consequence of the quantum laws. Thus, there is no problem with ordinary reality.

The realm of ordinary realism is considerable. Except for particle and atomic physics, including some fragments of chemistry, most of science deals with macroscopic objects and macroscopic parts of those objects. The same is true of biology, DNA and proteins being practically macroscopic. Some people have envisioned a possible role of quantum events in the mechanisms of life, perhaps in our brains, but although one may have strong suspicions against such speculations, this is not the place for discussing them. In any case, there are no cogent arguments for not granting that all of science, except the parts we mentioned, is perfectly classical and belongs to ordinary reality. It need not raise any philosophical qualms.

RATIONALITY VERSUS REALISM

When we deal with microscopic objects, complementarity forbids us a realistic approach. This renunciation is somewhat similar to Kant's rejection of realism when he declared the thing in itself, the noumenon, inaccessible to pure reason. Instead of the limits imposed on reason by categorical judgements, the constraints we cannot escape are now those of logic. Something real is necessarily something true.

However, a significant difference between reality and truth is that the former is existential and wordless, whereas the concept of truth is perfectly controlled by logic. This gives us a handle on the problem of realism. According to logic, true statements should obey some general conditions or axioms. The most interesting of these asserts that if a proposition a is true and another b is also true, then the proposition "a and b" should be true.

Most statements in quantum logic cannot be said to be true because of complementarity, even when they belong to a consistent family of histories and follow logically from a true fact. There exist, however, many statements that may be said to be reliable (or trustworthy, in d'Espagnat's terminology): they can be relied upon without fear of running into a logical contradiction. In a nutshell, the range of rationality is wider than that of realism.

Consider, for instance, a proposition *a* regarding a quantum particle at a given time. It belongs, together with its negation, to a consistent family of histories including all relevant phenomena (or, more plainly put, all the experimental data one could observe). Each experimental datum, a fact, is also stated as a property, and it turns out that one or several such data imply proposition *a* according to the laws of logic.

In ordinary reality, when a fact implies a statement, the statement is necessarily true. This is not so in the quantum world. It often happens that there exist several consistent families of histories with the same data, and in some of them another proposition *b* also follows logically from the data. If there is no consistent family including both propositions *a* and *b*, complementarity forbids us to consider *a* as true. Thus, the proposition "*a* and *b*" cannot even be stated, and of course it cannot be true. By sticking to a given consistent set of histories, there will never be any contradiction if one relies on proposition *a* "as if it were" true. This is what is meant by *a* being reliable or trustworthy.

THE "EPR" EXPERIMENT

No wonder such a tricky situation has given rise to innumerable discussions. The framework of consistent histories and their logical apparatus has contributed to its clarification, but in some sense it also made the situation worse with regard to realism, because everything is so neatly specified that there is no possible way out. In 1935, Einstein, Podolsky, and Rosen (abbreviated as EPR) proposed a way to introduce an element of realism in quantum mechanics. Their famous experiment is worth discussing. (We shall assume a certain familiarity with the notions involved on the part of the reader, since a complete explanation would be too lengthy.)

In a form given by David Bohm, the EPR experiment is the following: A particle Q decays into two spin-1/2 paricles P and P' in a state of total spin 0. The spin component of P along a direction n is measured at time t. Similarly, the spin component of P' along a direction n' is measured at a later time t'. The results of the measurements or, more properly, the corresponding data, are plain facts, and the questions we ask concern the spin of P' at a time immediately following t, when the spin of the other particle P has already been measured and P' itself has not yet entered the measuring device. What can be said?

According to EPR, the spin component of P' along the direction n, between times t and t', must be opposite to that measured for P. This is propositon a in our previous discussion. It belongs to a consistent logical framework in which it is implied by the measurement of P. In our terminology, proposition a is at least reliable. EPR considered it to be true, arguing that this property of P' is known without perturbing P' in any way. They called it "an element of reality," a glimpse of something real in the midst of quanta.

But such an element of reality cannot be true for the following reason: Consider another proposition b stating that the spin component of P' along the direction n' between times t and t' is already equal to the value that will later be measured at time t'. Whatever we had in favor of the truth of a still holds for b. It enters in a consistent logical framework in which it follows logically from the datum at time t'. Propositon b is just as reliable as a, and there is no logical framework, no consistent family of histories including both of them, so that proposition "a and b" is meaningless. None of them can be true, since they are on the same footing. EPR's element of reality has therefore no more reality than any other quantum proposition.

The above situation can be concretely illustrated as an argument between two persons. Consider then two physicists, the unavoidable pair Alice and Bob, each of them having made one of the two measurements. Then each of them can assert that she/he knows something about the spin of particle P' between times t and t'. "I know its component x"; "Myself, I know its component z." Since, according to quantum mechanics, these two statements are incompatible, they go on arguing. Each of them can argue that there is no possible logical flaw in his/her reasoning, and that logic is on his/her side. "I know how to think, my dear, and everybody in my lab can vouch for me." Neither of them can accept the other's point of view, since it is incompatible with hers/his. "Look, since it is perfectly clear that I'm right, you can't be." Frege, who knew the significance of a universe of discourse in logic, would have condemned them both for their ignorance, had he been the fourth judge in Hades.

The case when both directions n and n' are the same has been described in a humorous way by O. Steinmann. An interplanetary lottery is to be held on planet earth in a quantum fashion. An EPR pair of particles is produced on earth at time 0, particle P' being kept in a trap for the purpose of measuring its spin component along n at time t'. People play by betting on the result of this measurement. A cunning inhabitant of Saturn decides to cheat: particle P passing nearby, he measures surreptitiously its spin component along direction n at a time t before t'. He then knows with certainty what the result on earth will be, makes a bet, and, of course, he wins. This is real, Steinmann says, since what could be more real than getting the money? The lottery organizers suspect a swindling, but there is nothing they can do, because particle P grazed Saturn too late for a light signal sent from Saturn to reach earth (or the other way around) before the draw. The organizers, admirers of Einstein's, cannot claim that inside information was used and must pay up.

Is there a violation of relativistic constraints in this case? The answer is no, because the Saturnian had prior information, namely, how the pair of particles was produced at time 0, as well as information about a future event, namely, the direction n along which the measurement on earth would be made. This is essential, and the trick is that, although the result of each measurement is random, the two of them are strongly correlated. Some people find it hard to swallow that two particles separated by such an enormous distance could be so strongly correlated, but this is a fact of life. It has been experimentally confirmed and we shall now see how.

BELL AND ASPECT

Some readers may be wondering why nothing was said about the work of John Bell. It is now time to make up for this oversight, though once again we shall assume the reader to possess the necessary background for the sake of brevity.

John Bell was unhappy with the status of reality in quantum mechanics. Is there something real hidden behind quantum mechanics? If so, Bell made some very reasonable assumptions about those hidden features. Like everything else in ordinary physics, they should be describable by numbers, namely, by some hidden parameters. Since quantum measurements show random results, the hidden parameters associated with measured particles must be random but, being real, they must obey classical probability calculus, as does anything real that is not exactly known.

Bell considered the EPR experiment as we described it, with two directions n and n' as before. Time is not relevant in the present case, and we may take $t = t'$ (the two measurements being made on the two particles when they are far apart from each other and essentially at the same time). Let A (B) be the device measuring the spin component of particle P (P') along the direction n (n'). Bell made an assumption of *separability* between the two devices. He assumed that the result of the measurement of P by A depends deterministically upon the direction n and the hidden parameters for P as well as P', but nothing else. He made a similar assumption regarding the result of the measurement of P' by means of the device B. It should be emphasized that one explicitly assumes that the result coming from A does not depend on n', that is, that the two apparatus can ignore each other even if the two particles are classically correlated. As an example of correlation between real objects, think of two parts of a stable rocket after having been separated: if one of them spins in one direction, the other spins in the opposite direction. This is the kind of correlation that can be assumed.

From these assumptions, and using classical probability theory, Bell obtained in 1964 some inequalities for the combination of the results of both measurements, involving a few well-chosen directions n and n'. The beauty of his result is that these inequalities are not always satisfied by the predictions of quantum mechanics. The reason for the discrepancy is due to the quantum description of the two-particle state. This is a so-called entangled or nonseparable state, whose correlations cannot be properly represented by classical probability calculus. Bell's result thus opened the possibility of an experimental test for the existence of a specific, common sense kind of reality.

The experiment was carried out by several groups of researchers, the most precise results being obtained in 1982 by Alain Aspect and his team. The two particles were photons emitted by the same atom, and the spin measurements amounted to polarization measurements for the photons. The result was clear-cut on the side of pure, hard quantum mechanics against hidden separable reality.

We must say a few words about nonseparability. It means that the properties of a quantum system may happen to be correlated with some properties of another, faraway system. In the consistent histories approach, this means that, when the two systems are considered together with the measuring devices, consistency requires the right correlation between data. What is stated about the measurement of particle P is not arbitrary if a statement about the measurement of particle P' has already been chosen, at least when the two directions n and n' are parallel. This is a requirement of *quantum* logic. It goes against Bell's assumption, since the choice of the two directions, for well-separated devices, must be taken into account.

Much fuss has been made about nonseparability. For some people, it implies that quantum mechanics is holistic: it can speak only of the whole universe and not of its separate parts, even when one of them has no interaction with the rest of the world. It is as if a basic tenet of science, namely, the possibility of investigating an isolated part of the world by itself, were denied. Had these people been right, this would of course have been a severe criticism, since science is based on the study of limited objects, which is the mildest form of reductionism.

Such extreme conclusions are fortunately incorrect. One can perfectly well describe and use a system that is sufficiently isolated from the rest of the world. This system may involve as many experimental devices as necessary. Nonseparability only implies that two such systems, with no direct mutual interaction, can show correlations in the results of their respective measurements in some special cases. These cases can always be determined by a careful consideration of the preparation device and, at any rate, any *fact* observed in one system is not changed because of the existence of the other system. Measurements are correlated, but who cares? This is not a direct influence.

Mathematically, nonseparability amounts to the fact that a wave function for several particles is not generally a product of individual wave functions for each particle. This is particularly the case for identical particles, electrons or photons, for instance, for which the global wave function must be antisymmetric or symmetric according to Pauli's principle. Nonseparability thus ranks among the deepest principles of quantum theory, and its benefits far outweigh the slight philosophical qualms it may have provoked.

Nonseparability, or Pauli's principle for electrons, explains why a table made of wood or steel is hard, why atoms can bind in a molecule, why matter is stable and does not collapse into nothingness, as well as many other effects, too numerous to be listed. Nonseparability of photons is, on the other hand, necessary for a laser to work. Those who prefer nature to be separable should therefore stay away from nightclubs. Pity on them; were they to have their way, it would be a proof of their own nonexistence.

As a final comment, Bell's assumptions seem so reasonable at first sight because they are correct for the description of random *classical* events. Their classical validity may be proved by the same techniques used to recover common sense from quantum physics, and they appear reasonable because they belong to common sense. If they failed under an experimental test it is simply because common sense cannot be extended to a genuine quantum system. That's all.

CONTROVERSIES ABOUT HISTORIES

Some readers may have heard that consistent histories have been criticized in the physics literature and, since a large part of this book makes use of this approach, the question should be clearly restated.

Maybe a brief history of the main events might help to bring them into focus. Consistent histories were first proposed by Robert Griffiths in 1984 and their logical background was noticed by the author in 1988. Two years later, Murray Gell-Mann and James Hartle reformulated them in accordance with the notion of decoherence. Criticism came first from Bernard d'Espagnat in a somewhat roundabout way. He argued that readers of these papers might get the impression that consistent histories restored naive realism in the quantum world—without asserting that the authors had claimed such a thing, because they had not. As a matter of fact, we were at the time too busy exploring the consequences of the new theory to worry about philosophical issues.

D'Espagnat was nevertheless right in reminding us that such issues are important. For reasons that should be obvious from our earlier discussion, he raised the question of truth. "What should be said to be true within the history framework?" he asked. And he carried on his attack by pointing to a weak link: he noticed that Griffiths had perhaps a bit hastily used the word "true" in some places, while Omnès had been mischieviously cautious in never employing it.

In trying to solve the problem, I was led to formulate the notion of logically reliable though not true propositions, a notion endorsed by d'Espagnat himself, who called these propositions trustworthy (and of course not entirely true). Unfortunately, having too hastily worked out the solution of a problem that had been forced on me, I proposed a criterion for defining true properties in quantum mechanics. It appeared to be a mild one, since the true statements it allowed, besides facts, were only the results of quantum measurements and the classical properties of macroscopic objects when these are not under observation.

The criterion I had proposed was wrong, as Fay Dowker and Adrian Kent have shown. Their conclusions somewhat exceeded their results because they worked only with algebraic techniques, which did not allow them to take a full account of the decoherence effect and the peculiarities of classical statements. They had nevertheless a good point and I willingly retract the said criterion.

What can be said to be true? Facts, of course, are true, but what else? The essential obstacle is complementarity, or the multiplicity of consistent logical frameworks describing a microscopic system. Truth must be immune against the ambiguities of complementarity. Moreover, when a true proposition is added to the facts and logically follows from them, no other proposition incompatible with it should be possible. The conclusions concerning reliable "non-true though not untrue" propositions then remain correct, but everybody already agreed on that point.

The classical properties of a macroscopic object that is not under observation are still true within the framework of classical propositions (derived from quantum theory). We should add that no quantum measurement or anything like it must be taking place, but this can be expressed as a condition on the preparing process in terms of histories.

What about genuine quantum propositions? Only very few of them can be said to be true. They do not even include the result of a measurement as a real property of the measured system at the time it was measured, except in a few cases. Truth is mostly reduced to a definite property of the measured system resulting from the measuring process immediately after complete decoherence in the measuring device. Even that must be qualified: the measurement must be an "ideal" one, which does not spoil a so-called eigenstate of the measured observable. To put it another way, an ideal measurement gives the same result twice if performed in immediate succession. It might appear as a meager conclusion, were it not for the fact that it covers the cases where the rule of wave function reduction has been used for so many years in the practice of physics or, in other words, the essentials of the Copenhagen interpretation of measurements.

The conclusion is therefore that not much can be said to be true in the quantum world of individual events. Reality remains veiled, in d'Espagnat's words. The little that is true, or real, is, however, sufficient for doing physics, if one insists on introducing the word "true." In fact, the word need not belong to the vocabulary of physics, except for facts.

Does this mean that histories have suffered a severe blow for not succeeding in reaching realism? Not at all: realism was not their goal, and they have no claim on a restoration of some naive realism. What are they, then? The answer is simple: a method.[1] Histories were used as a pedagogical method when the archangel taught the young angel physics. As for human physics, histories provide a method for putting order in a subject—interpretation—that would otherwise easily turn into a labyrinth, as it often did in the past. This method introduces logic in a subject much in need of it. It is a method for proving: it adds nothing to the basic principles of the theory, each of them separately confirmed by experiments, and only makes use of these principles, just as every method is supposed to do.

No method can claim a monopoly on correctness, for the same

---

[1] I am here expressing a rather reductive view of the power of histories, which has the merit of avoiding controversy. Perhaps histories have more power in store, which could be used in quantum cosmology, for instance. Only time will tell. It should be clear that the restricted value I place here on histories does not bind others working in the field who may expect more from them.

results may be obtained with other methods. Only the principles on which it stands and the conclusions it reaches are important. Does it matter if a theorem is proved by an algebraic or an analytical method as long as it is proved? Till now, no method has had the scope or obtained some of the basic proofs that were achieved with the method of consistent theories. Other methods may very well do, only to reach (necessarily) the same conclusions.

Many of these conclusions, by the way, had already been guessed by Bohr thanks to a stroke of genius, unaided by the powerful mathematical techniques and the guide of a discursive construction available to latecomers. Or perhaps he found them through a lifelong meditation worth of our admiration. The method of histories allows one to prove rather than guess or gloss endlessly over Bohr's writings. It turned interpretation into an ordinary discursive theory anyone can check. It also showed that the questionable railings against tomfoolery Bohr had to erect, such as his drastic separation between quantum and classical physics, are not necessary, and their removal greatly opens up our vistas.

The only comment we wish to add concerns the choice of the properties entering into the practical use of histories. This choice was criticized because of its arbitrariness, but it can be very easily justified: a physicist needs to describe what he is doing in simple words. She needs to draw conclusions from observations with the help of logic. She wishes to put some of the steps in a form that can be directly investigated using the theory's mathematical techniques. Which properties to choose? Only those most convenient for the purpose at hand. Many other descriptions may do as well, all different because of complementarity. Some of them will lead to the same conclusions and they are just as good. Others will be useless, not necessarily wrong but only idle talk of no consequence. Why bother? Asking questions about the existence of useless histories amounts to performing calculations that are of no help in solving a problem. They belong in the waste-paper basket.

TOWARD A WIDER REALISM

What are we to conclude? Realism, in so far as it recognizes that the world does not depend on us for its existence, is unassailable. When it says that the ever-changing world our senses perceive is real, this is just a definition, the definition of the recognition of "that." In saying that ordinary reality agrees with common sense, one is simply stating an immemorial observation. Quantum mechanics adds only that its laws have no objection against such an observation and, in some sense, make it deeper, in harmony with universal laws. What a wonder this was for the pre-Socratic philosophers we met in Hades.

However, our words, our vision, and many familiar philosophical principles humans drew too hastily from common sense, break down when confronted with the atomic world. The laws of this world are in a sense real, since their consequences have always been found to be right. Have we reached their ultimate form? or do they have other facets, new extensions we have not yet found? Whatever the case, there is no reason to expect wider laws to be less formal than the present ones. We are in any case left with the cogent philosophical task of coming to terms with formal science and mastering its meaning.

Our next task will be to attempt an approach toward a wider realism, fearlessly confronting formal science and mathematics together. This is the birealism we shall propose in the last chapter of this book, maybe too bold a step for a physicist. Bertrand Russell once said that there are no worse philosophical books that those written by scientists seized by the middle-aged love of philosophy. It is also often said that science by itself cannot produce any new results in philosophy, only decide whether certain philosophical statements are valid or not, and this is certainly true for a definite branch of science. The very existence of science, however, its degree of universality, and some of its characteristics, raise obvious questions of a philosophical nature, for which science, as a total object of reflection, can suggest tentative answers.

We shall not go very far toward the wide or grand realism offering its vision now in a remote perspective. Only some possible trails, maybe, will be sketched. They would be much better followed by true philosophers, and my only aim will be to offer them a few hints for a long and fascinating journey, and to invite you, the reader, to the joy of contemplation.

PART FOUR

STATE OF THE QUESTION

AND PERSPECTIVES

✢

A New Beginning

A PRELIMINARY REPORT

IT HAS BEEN a long journey, even if we have taken some shortcuts; but it is not over yet, and we have every reason to go still farther. We cannot ignore the signs along the way pointing to the almost immediate presence of a new philosophy of knowledge.

Let us begin by reviewing the situation. We have started from a state of knowledge familiar to everyone and whose broad lines can be retraced. It is first of all an existential situation where humankind penetrates time and space, and matter too. Humanity is aware of the extent of the universe, it probes its birth and reconstructs its history; it knows the unity that transcends life's diversity and also knows its place in it. It is an intellectual situation as well, where we possess a science that is incomplete, to be sure, but how revealing! A science that has shown us the existence of very profound laws at the heart of things, laws that are not discordant but harmoniously meshed into a tight bundle. A science that has also revealed a coherence between the products of our intellect and the outside world, between Logos and Reality, in other words, between the major terms in yesterday's and in today's philosophy.

And yet, this science appeared obscure and inscrutable, with its heart caught in the thick thorns of its formalism. It was by ploughing through that thorny tangle that we could see the view change. A peculiar science, quantum mechanics, came to the rescue, no doubt because it is the one that can most deeply penetrate the bundle of laws, perhaps down to its very origin.

We have learned some extraordinary things, even if not all of them are equally convincing. For the philosopher, the most important consequence is the reversal of an intellectual approach several times millenary, which we presently follow in the opposite direction. Surely, we agree with Hume that it is the world around us which, through our senses, forges our thought structures; first on an individual basis, and later spreading to the community of humans by means of language. The evolution of our species and of those that preceded it took place in this world, by yielding to its secret order, concealed but persistent, and gradually refining our perception. But we reject Hume when he declares that the source of the world's order is inaccessible, and we also reject Kant, for he sees that source only inside ourselves. The source is out there, in the laws we now know well, or at least sufficiently well. We know enough of the world's order to rely on it.

Thus, to start from common sense alone is out of the question. It is by reflecting on the nature of this common sense that in the past philosophy chose its principles, declared them unassailable, and drew up their list. Then, based on them, it believed that it could pierce through everything the mind could think. But those principles collapsed one after the other when they were confronted with the world of the "infinitely small": intelligibility (or the possibility of representing reality in our minds), locality (each thing has a place of its own), causality (every effect has a cause), discernability (two things that are not the same can be distinguished by the mind), cognizability (if an idea concerning the world can be thought, it can in principle be decided whether it is true or false). Philosophy's dream of explaining the world was vain, at least in the sense of its own idea of an explanation: to have a clear picture in the mind of the thing being explained, an image that could be put into words and communicated to others through those words. It is to symbols that we must now resort.

But those symbols contain the concepts and express the laws like principles of another kind, and we have seen how the reversal takes place. Once these new principles have been conquered, through painful efforts and lengthy reflections, they can restore the world. It is in them, impregnated with matter, and not in our mind that the source of logic, and hence of reason, lies. Our vision of the world with all its appearances takes roots on these principles and emerges again as one of their manifestations. We no longer attain the principles of the world through the ordinary language of reason, but instead obtain an incomparably stronger consistency by deducing reason from those principles.

One more revelation must be borne in mind, or at least examined closely if one still has doubts about it: the chasm, as we have called it, the unbridgeable gap between theory and the real world, between thought and existence, or, to use our previous terms, between Logos and Reality.

Such is the new state of affairs that we must now face.

THE BEGINNINGS OF A PHILOSOPHY

Recall Bacon's words, which we have perhaps stretched beyond their author's intentions to make them sound like a prophesy: the most general axioms of science will be reached only at the end, and it will then be seen that these are not deceptive notions but well-defined concepts that Nature will recognize as its first principles, present in the heart and the essence of things. Would it be possible that today those words could be spoken in the present tense, and that science could be ready to give birth to a new philosophy?

This question is more than a simple suggestion, and what we have just said regarding quantum physics almost compels us to go ahead. This science, so singular and revealing, contained in its principles the instruments of its own interpretation. Likewise, could it be that all of science should lie so close to the heart and the essence of things as to give rise to its own philosophy? A philosophy of knowledge, to be sure, but isn't it a prerequisite for any philosophical enterprise—other than doubt, of course?

Science has come a long way. It has traveled from reason to the absolute symbols of mathematics, and from ordinary objects to their universal laws. In the beginning, ignorance and darkness permeated everything; from language itself, the instrument of reason, to the objects surrounding us. True, the latter concealed their mystery and appeared at first as obvious, as irreducible. We now see them differently, better in fact, and the sources of reason also begin to show up: in a world where order is pervasive, a life born out of a chemical process that evolves, with increasing complexity and strength, right to the human brain, the organ that perceives the order. Many links are still missing, and the culmination of the whole process, our brain, is only beginning to reveal itself, but a few broad outlines may be perceived already.

Thus we see science start from the unknown, and from this darkness attain the point where the beginning, previously accepted without questioning, is all light. Science reverts to its own origin, like a circle destined perhaps to be perfect. But such a circle, even if complete, will still be a circle, without beginning or end, that is, without its own guiding principle, in other words, without philosophy. This is why we must break it, so it can bear fruit.

To break the circle means finding that which it cannot learn about itself by itself. It is finding a founding principle for science that science itself cannot provide. Only then, perhaps, may metaphysics begin.

Science developed in opposition to metaphysics, and it had to. There was even a time (which many believe includes the present) when metaphysics was thought to be dead, wiped out forever. Hume ridiculed it, trampled it, but had to impose a proscription that was itself metaphysical: the absolute impossibility of reaching the source of the intrinsic order of things. We now know that, in that respect, he was totally wrong.

What do we mean here by metaphysics? We know its etymology: "that which comes after physics." According to the scholars, the name might not have the deep meaning we would expect, but originated in connection with a catalog. Aristotle's books did not have titles (not all of them did, at any rate). One of them was called *Physics*, the same name ("*On Nature*") given by so many earlier authors. The book next to it on the shelf would have been given as title *Metaphysics*, "coming after physics," the next one. I would rather adopt the sense this word appears to have: the result of reflections provoked by a certain knowledge of *physis*, of nature. I would also add the sense of the prefix *meta*, the same one "metalanguage" has in logic: a way to penetrate that which cannot be sufficiently exploited in its own language. To sum up, it is all about trying to learn, trying to reach the things science carries within it but cannot itself tell.

Thus, I claim that science is presently mature enough to permit the revival of metaphysics. Obviously, such a claim cannot be the consequence of a proof but, at best, of a conviction. It is also an expression of hope, of encouragement, addressed to the philosophers of the future, who will contemplate with indulgence the dereliction of those of the present. To talk about this new enterprise I will gladly borrow Bacon's words about science: This instauration will be by no means forgetful of the conditions of mortality and humanity, for it does not suppose that the work can be altogether completed within one generation, but provides for its being taken up by another.

This was perhaps the most I could say, because from this point on the journey is so full of surprises and possibilities that it would be impossible to anticipate them. I will nevertheless carry on, warning the reader not to see in what follows anything more than an outline, the uncertain contours of incipient ideas.

It would be advisable to stop here as well our discussion of the state of the question, because I find more confusing than enlightening the contributions of certain authors of contemporary epistemology, some of them among the most popular. This remark does not of course apply to the rich and necessary works of historians, nor to some older books worthy of serious consideration, even if the state of science at the time they were written renders them partially or totally obsolete. Notable among these are the books by Bachelard in which he provides a touch one would like so much to find elsewhere, and which I have tried without success to introduce here: the touch of the poet, the only one that remains, together with that of the visionary, in a passing knowledge.

The Religious Temptation and the Sacred

I would like to close this chapter on a question that many readers might probably have raised themselves: the connection between the above enterprise and religion. There are many books these days whose authors believe they have found in science the signs of the existence of God. Christians (and no doubt Jews as well) see in the theory of the Big Bang the confirmation of the story told in Genesis. Others, or the same ones, assimilate the Law of the Old Testament, or Torah, to the laws of nature. Yet others cite some oriental religion, such as Tao-tö-king, as evidence. It is true, as the last example shows well, that those texts are highly poetic and therefore allow for a rather fuzzy interpretation. Besides, our modern authors' reading of them is entirely based on analogies. We must however ask ourselves whether there is here something other than a mere play on the ambiguity of words, even if our authors take themselves very seriously. But are they really?

On tackling this kind of subject, even briefly, one must clearly show his true colors. Thus, I, the author, call myself a Christian, though my preferences in matters of belief are closer to Nicholas de Cues' *Docta Ignorantia* than to Thomas Aquinas' *Summa Theologica* or Karl Barth's *Dogmatik*. By this personal note I wished to assure my Christian friends that the targets of my criticism are only certain thoughtless proselytes.

I shall bring up only one argument, too often repeated: the interpretation of an astonishing scientific discovery—the probable, if not certain, existence of a beginning of the universe—as a proof of the creation of the world and, as a consequence, of the existence of a Creator. This is a flagrant lack of logic. Let us take a closer look. Within the framework of general relativity, stretched to its limits, there is one particular solution to Einstein's equations that appears to be by far the most plausible. It implies a so-called homogeneous and isotropic universe (that is, having the same properties in like degree in all directions), in agreement with the observed distribution of the galaxies and, especially, with the thermal radiation that currently fills the universe. The special solution so obtained provides a mathematical model of the universe and its history presenting what is known as a singularity, a barrier that the laws of physics forbid crossing, located in the past and beyond which time cannot be extended. Delving into the model thanks to our knowledge of the laws of physics, we can derive various consequences: the present amount of helium and of other light nuclei, all the characteristics of the thermal radiation mentioned above, and Hubble's law on the recession of galaxies. All these consequences are reasonably well confirmed by experience and they therefore lend a high degree of plausibility to the model. One can then logically believe in this model, and much of it is likely to be true. Let us assume so.

What have we proved? Something extremely important for the physicist: the fact that the laws we discovered here now apply to the whole universe. But then, what has God to do with all this? Do we need him as the Creator? This would amount to imagining a limitless time containing at some point the instant of creation. We can conceive some Jewish author writing such a story at around Ezra's time, but today? The science on which the argument is based, general relativity, is unequivocal on this point: no physicist can give any meaning to the notion of time beyond the singularity, beyond the "beginning." Moreover, Saint Augustine came to the same conclusion in his *Confessions*. To those who wondered,

"What was God doing before he created the world?" he replied that "before the world existed, time did not exist."

We all have mental pictures, Kant's famous forms of sensible intuition. We cannot "imagine" a reality that is not a content in a container, or does not extend to infinity; this is as true of time as of everything else. In our imagination, a barrier or boundary comes with the other side of the boundary, with an exterior, and the exterior of a universe with a bounded past comes with the most imprecise image of all: that of a God the Creator. Those who see it in this way must eventually identify it with their most inner feelings, and we then speak of a mystery. Indeed, the identification of the perfect exterior with the perfect interior is a mystery, but it is also called a paralogism (i.e., faulty reasoning).

The logical inconsistency goes even further. By assuming the existence of a Creator, one is really looking for a cause; supposing that this Creator existed before the universe did amounts to seeing in the beginning of time only a particular stage of a larger story. One must not forget, however, that the laws of physics within whose framework these facts take place have taught us something else. The notion of cause is not absolute. Space and time must be conceived in themselves, without any external container: it is one of the starting points of the theory from which the model was constructed. Thus, it is all too easy to stick onto what we know concepts that contradict the assumptions on which they are based. This is another paralogism.

Finally, the idea of a God the Creator resuscitates an ancestral image but sidesteps the true mystery: that of the immanence of laws. It is they that create this universe, or at least structure it through its extension in time. Does this mean that all God has to do is create laws? But then, what does the concept of God add to the concept of law? A cause? Come on, this would be giving in to mental tics! The laws, by their very universality, are completely impervious to what is external to them. They should be our first object of meditation, for we can construe them in direct connection with everything to which we have access. The domain of religion is elsewhere than in the creation of the world.

And yet, some might reply, didn't Einstein say, "The conviction that the world is governed by rational rules and can be apprehended by reason belongs to the domain of religion. I cannot

243

conceive a true scientist lacking this profound belief. The situation may be expressed through an image: science without religion is lame, religion without science is blind"?

It seems to me that we must distinguish here between two words. Einstein's reflection acquires its full significance if "religion" is understood in the sense of "sacred." The latter captures a concept that Mircea Eliade introduces in the foreword of his *Histoire des croyances et des ideés religieuses* (*A History of Religious Beliefs and Ideas*) as follows: "It is difficult to imagine how the human mind could operate without the conviction that there is something irreducibly *real*[1] in the world; and it is impossible to imagine how consciousness could appear without conferring a *meaning* on human impulses and experiences. Consciousness of a real and meaningful world is intimately connected with the discovery of the sacred. By experiencing the sacred, the human mind has grasped the difference between real, powerful, rich and meaningful things and others not possessing those attributes, that is, the chaotic and dangerous flow of events, their haphazard and meaningless occurrence and disappearance. . . . In short, the 'sacred' is an element in the structure of consciousness, and not a mere stage in the development of that consciousness."

If we compare this conception of the sacred to the definition given in a dictionary (the French dictionary *Robert*, in this case) we notice a similarity: something is sacred that "merits an absolute respect, which may be considered as an absolute value." This is quite different from another common denotation: that which is sacred "belongs to a separate domain, forbidden and inviolable (as opposed to what is profane), and inspires a sentiment of religious reverence." This second meaning is accompanied by references to words such as "saint" and "taboo." It is preferable to exclude this latter sense opposing sacred to profane, since it establishes a duality clearly absent from Einstein's idea.

Mircea Eliade actually defines sacred twice in the text cited above, and his two definitions are different: he considers it first as something powerful and significant in itself, and later as a way to experience this power by a particular disposition of consciousness that he regards as a structure of the latter. We are not in a position to decide whether the sacred is a structure of consciousness or a

---

[1] The author's italics.

cultural inclination; it would be enough to admit that it is a state of consciousness that many, if not all, of us know under some form or another. The important point is to grant that the sacred is a disposition experienced by the individual, and that it therefore establishes a relationship between the world and human behavior or—why not?—between a philosophy of knowledge and humanity.

Hence, it is the first quality that Eliade attributes to the sacred that is for us the most important: the quality of being powerful, rich, and meaningful. We may also observe that some of Eliade's reservations are unnecessary. When he talks about "the chaotic and dangerous flow of events, their haphazard and meaningless occurrence and disappearance," he seems to assume that this domain, repelling to the sacred, may belong to some primal reality independent of any form of order. Now, we know that such a disorder is only apparent: ill-fated circumstances or a tragic accident may appear fearsome or fatal, but they are nevertheless governed by a higher order, one closer to the laws. The flow of things may be dangerous and loaded with risks for the individual, the group, or even the species, but there is nothing chaotic in its mechanism, even though it remains complex and unpredictable. The appearance and disappearance of things may seem to occur at random, but they are never senseless. To sum up: the way we see it, the sacred is everywhere in the universe and nothing is completely profane. Profanity is but an illusion of our own ignorance, the slumber of the mind or the madness of our false ideas.

What Is Science?

$\mathbf{I}$N THE PRESENT CHAPTER we proceed with our review of the state of the question by inquiring into the nature of science.[1]

SCIENCE AS REPRESENTATION

Every thought rests on some representation. This is how the mind translates our perception of the world. Our memories of it are probably located in the circuits of neural signals that develop under the effect of repeated or violent perceptions, and later become fixed. We perceive a landscape as a whole, vast and still, but at each instant our eye catches only a minute part of it; it is in our memory that we contemplate the picture that emerges out of thousands of those fleeting impressions: representation. Even our words serve to represent.

Thus, to the question "What is science?" we shall answer that it too is a representation of reality. Not the primary representation imagined by Locke and Hume, constructed from pieces coming directly from reality, but rather an abstract and coded picture, albeit a faithful one.

Humans possess a variety of representations of reality: magical, poetic, ideological, and others still. These live in a philosophical system, a religion or a culture, sometimes simply in a state of mind. Each of them has its own language and, conversely, our language is made up of bits of shattered representations, ready to be combined to produce yet other shifting representations. What then makes science distinct from all that? Is it because it employs its own concepts, inspired by experience, or because science is unique

[1] We shall employ the term "science" to designate what are usually called the physical sciences: the study of matter, the celestial bodies (including the earth), and living organisms. Logic and mathematics will retain their respective names, thus marking their unbridgeable distance from concrete reality. We are aware that the place of the social sciences then remains ambiguous, but we do not need to discuss them here anyway.

by the rigor of its arguments and its logic? Not quite. The first trait applies just as well to Pico de la Mirandola's hermetics and the second to scholastic theology. Might the answer lie in the presence of the laws? This is unlikely, laws being part of every vision of the world.

How about logical consistency? Theology too strives for consistency, and it was the source of it, in connection with questions on the nature of the gods posed by the ancient philosophies. Science did not possess it in those early times when it was only empirical and busy gathering facts. It came to it later, with maturity, as its different parts moved closer together and merged. We may nevertheless say without exaggeration that complete logical consistency has become a major attribute of science, always ready to be put to the test even at the risk of losing it all—something no theology would consent to do.

Indeed, this consistency is perpetually under scrutiny and questioning. Scientists spare no effort to track down eventual contradictions, and they constantly test the limits of their knowledge. Contrary to what some believe when they speak of "official" science's self-importance, the scientific community highly values the uncovering of an inconsistency, sometimes even more than a new discovery. I would not like to paint the situation in idyllic colors either. There are too many examples where the obstinacy of scientists rivaled that of ordinary mortals—it suffices to recall the fierce denial of Wegener's continental drift. What I have just said is therefore valid in the long run and not always in the immediate present.

Unlike other alternative or competing representations, science demands absolute consistency. A single clear inconsistency renders any scientific branch suspect and untrustworthy. Should it persist for too long gangrene may set in and spread to the entire body of science. The expectations for consistency are now so high and uncompromising that to live up to them science must be ready to make atonement by offering itself as a sacrifice.

There were many examples of this in the course of history, as when a contradiction appeared between instantaneous gravitational force and the impossibility of actions faster than the speed of light. Another time it was the collapse of Rutherford's classical atomic model that claimed an exemplary sacrifice on the altar of logical consistency: that of intuition and common sense.

To be sure, these sacrifices were not in vain, and after each alarm science recovered its beautiful consistency, more confident than ever. Such catastrophes invariably followed by a spectacular rescue end up by dissipating any fear, and physicists now prize the discovery of the slightest inconsistency. They search for it and track it down, for they anticipate a major breakthrough more than they fear a real danger. And yet, despite this almost absolute confidence, faith in science stems above all from a greatness comparable to that of a naked warrior: to prevail by accepting its own vulnerability.

On Certain Types of Laws

Science represents the world as bundled up inside a tight network of laws. These rules or laws have a tremendous significance, but it is difficult to crack their essential nature; we can only acknowledge them and recognize their enduring, pervasive action.

There are in fact different categories of laws, and even if there is no widespread agreement on their names it is convenient to distinguish three types: empirical rules, principles, and laws. Among the numberless empirical rules, there are first those we might call primary. They come from events that repeat themselves endlessly. Leaves turn yellow in the fall, the sun appears red at sunset, cats have whiskers, orange skin is of a certain color; all those things form a loose collection of primary rules, resulting from the repetition of things that weave our visual representation and language.

Science often starts from an attentive analysis of such rules. It is in some sense what Linné does when he specifies the multiple similarities and differences that exist in the vegetable kingdom. In so doing he obtains more elaborate, secondary empirical rules, the only ones we shall consider from now on. These often take a quantitative form: Ptolemy's rule of epicycles for the planetary motions, Kepler's rules for the same phenomenon, Ohm's "law" in electricity, and many others. Each of them remains nevertheless more an observation than an explanation, a summary of observed facts that accounts for their appearance but cannot pretend to go beyond that.

Principles are altogether different, and their ambition stands no comparison with the modesty of empirical rules. A principle must be universal. The idea came from Greek philosophy and through medieval theology, but it is in physics that it found its precise and exclusive meaning. Although biology also has its principles, they contain a larger degree of vagueness. Thus, evolution is a great principle, but its statement leaves ample room for interpretation. We have seen that the first science claiming to have principles was Newton's mechanics, and have already mentioned how he intended in this way to free us of our terrestrial condition. But he nonetheless refused to see in his principles anything besides a summary of facts and experiences, an empirical rule of a higher, more economical order. We must, however, single them out, precisely because they are universal and may therefore be used to make predictions.

To pretend that a principle applies universally may appear to be a hopeless enterprise, like betting against an infinitely rich bank whose subtleties are unpredictable: the bank of Reality. Indeed, if principles are universal, they must account for every phenomenon falling within their jurisdiction with no exception (and subject to stringent quantitative constraints). This also holds for experiments that have never been performed or even imagined. It was far from obvious that Newton's laws together with the earth's rotation entail the rotation of the plane of a pendulum, and Foucault's followed by more than a century the principle that predicted the nature of its motion. A principle's universality thus applies to the unknown as well as to what already exists; this is both its strength and its vulnerability.

The above requirement is extremely severe, because a single discrepancy between reality and the expected consequences of a principle would signify the latter's demise. But at the same time, what a victory when the principle frustrates reality's blind attacks one after the other. Olé! In the cosmic arena, the principle is the bullfighter and matter is the charging bull!

Finally, after the principles come the laws. We understand by laws those particular consequences that can be deduced from the principles and which apply to some specified category of phenomena. For example, Kepler's rules have long ceased to be merely empirical to become a direct consequence of Newton's principles; thus, they are no longer empirical rules but have reached law status. Laws may therefore be seen as the principles' children, their offspring, as well as the means by which principles can be tested.

249

How do we know that a principle has prevailed and can be trusted? Only by having verified all its conceivable consequences through experiments, as far as this was possible. In the case of classical mechanics, such a verification period lasted almost two centuries and is still going on, albeit at a slower pace. We may almost say that certain laws are checked twice: first as theoretical consequences of the principles and later through experimental testing, which transforms the laws also into empirical rules. When the principles of a science are discovered, many empirical rules may change status and become laws if it is possible to derive them (theoretically) from the principles, as was the case with Kepler's rules/laws.

The consistency of contemporary science, in particular the physical sciences, may be measured by the fact that the number of laws greatly exceeds that of purely empirical rules. The existence of empirical rules that cannot be connected to the principles may indicate that the latter are still incomplete. Take, for instance, the empirical rule of springs that relates the elongation to the force of traction. Newton's mechanics could not make it a law for lack of a satisfactory explanation. In hindsight, it was a feeble signal pointing to the atomic structure of metals and the underlying existence of quantum principles.

The Transformations of Science

As is the case with any other type of representation, science progresses. This is a consequence of its human component, which is at the mercy of history, but it poses a serious problem, for the principles too might by called into question despite their claim to universality.

Let us consider genetics, for example. Mendel postulated the existence of genes as a principle that transcended the empirical rules of heredity. These genes, carried by the parents, were transmitted to their offspring according to the laws of probability. The discovery of chromosomes supplied genes with a concrete support, and showed that chance enters the picture when meiosis occurs, that is, at the time the first cell of the offspring is formed out of those of the parents. Mendel's "principle" was thus reduced to an empirical rule: that of the observed behavior of cells and their modes of transformation. New principles would later appear with the discovery of DNA and its replication rules. But are these true principles or simply rules? It is hard to say, since genetics, like all other life sciences for that matter, is in an intermediate state, not yet resting solely on universal principles. Besides, it is possible that these sciences only need principles that are highly plausible—that is, rules that are frequently verified—but not absolutely certain.

Physics is a different story, if only because of its ambitious claim to universal principles. We know that such an attitude almost resulted in its collapse at least three times in the past, and we have already related those episodes that marked the emergence of a new science: special relativity, followed by the relativistic theory of gravitation, and finally quantum mechanics. Each time, the new science swallowed the old one, it fed on its substance and restored it under a form only barely different. For instance, the principles of Newton's mechanics became particular laws of relativistic quantum mechanics, that is, consequences of more general principles. Unlike principles, whose universality is by definition absolute, most laws have a precise domain of application, determined by the hypotheses used to deduce them from the principles. Thus, when the demoted principles of classical mechanics became simple laws, they also saw their domain of application restricted to certain phenomena: those whose velocity is small compared to the velocity of light, and where Planck's constant is too small to play a significant role.

And so, strangely, the historical evolution of science seems to confirm the existence of universal principles, or at least strengthen our confidence in their existence. It is also a call for caution, for it suggests that today's principles are perhaps merely the reflection of others still unknown. At any rate, it would be a mistake to adopt a simplistic view that would reduce science to the temporary scale of our human values, something whose nature changes with time, where yesterday's certainties are simply the outdated beliefs of a bygone era. The revision of principles we have just mentioned has led some philosophers to say that the laws of science are vulnerable, changing with each new discovery and even with the spirit of the time, perpetually trying to keep up. This is to ignore the constant and watchful presence of Reality.

The above remark is worth elaborating on, because too often misunderstood. Some critics focus more on the words employed at a given time to state the principles than on their formal, mathematical structure. Once again, the slight attention paid to formal structure is a source of serious misunderstandings. If certain principles have disappeared as such, one cannot overemphasize the fact that they have become laws, and that this change in status resulted from the discovery of other principles, more general than the previous ones. We should not be impressed by this mutation, but rather reflect on the following wonder: every time science has offered itself as a sacrifice, it has risen to new heights instead of perishing, and has attained a higher degree of universality. Such episodes do not resemble the erratic course of human history but carry the unmistakable sign of the ringmaster: Reality and its supreme order, of which science is merely the servant and the scribe.

Thomas Kuhn

It is impossible to discuss the evolution of science without citing Thomas Kuhn and his most famous book, *The Structure of Scientific Revolutions* (1962). In it he proposes two main theses, one of which is precisely the existence of certain transformations in science that he calls "revolutions." The term is certainly excessive, as he later recognized, but it properly conveys the magnitude of the tremors that occasionally affect the scientific representation of the world.

His other major idea is his preference for paradigms over principles. In his opinion, a momentous discovery has a greater impact on the course of science due to the example it offers than by the principles into which it can be condensed. The breakthrough then constitutes a model to be imitated, a reference to be used as the basis for new research, that is, a paradigm (a word that was hitherto employed principally in grammatical analysis to indicate an example or pattern that serves as a model for many others). Thus, when Euler applies Newton's method to fluid mechanics he is attaching more significance to the success of Newton's mechanics than to the strict application to fluid masses of the principles stated by the latter.

The comparison between paradigms and principles is not, however, very relevant to our main argument. Indeed, the difference between imitating a paradigm and applying a principle seems to be a question in the psychology of researchers, a domain which I do not intend to study here. What counts for our purposes is the judgment passed on a new discovery: Does it confirm or contradict the principles we already know? In fact, Kuhn's thesis, whose interest for the study of history is undeniable, deserves to be put into perspective in order to better take into account two major transformations of science that he failed to consider: the arrival of formalism and the rise of consistency. It appears to me that both these events, which are not revolutionary because they evolve in time rather than occuring suddenly, are better understood within the framework of principles than by the dialectic of paradigms.

It was from that angle that we highlighted earlier the historical importance of the formal approach and its manifestations through Maxwell's equations. Now, if those equations have often inspired new research, they did not appear to have served as a paradigm because of their formal character but for some other feature. The rise of formalism, first in relativity and later in quantum physics, does not seem to have been inspired by a paradigm either, but rather dictated by necessity. Hence, the emergence of one of the most important characteristics of science occurred too gradually to be called a revolution and was not the result of imitating any paradigm.

Kuhn tends to link his two theses a bit too much, as if every "revolution" should necessarily be accompanied by a new paradigm. The continuity of scientific evolution then appears divided up into neat episodes, like a television series. Mendel's genes and the double helix structure of DNA discovered by Crick and Watson are two paradigms and the starting points of two revolutions. But their continuity is certainly much more important than their disparity.

Nonetheless, the term "revolution" perfectly describes certain specific events, such as the three transformations mentioned above that resulted in the birth of relativity, the relativistic theory of gravitation, and quantum physics. Each one of them was a true crisis, which science might not have survived. However, the important thing is not the crisis but its outcome: new, highly formal principles. This is something that Kuhn could not see through the prism of his paradigms, for there was no shortage of paradigms at the time, several every year; fireworks, fleeting, changing reflections that dazzle the eye. Can someone watching them really see?

253

If we focus not just on paradigm shifts but on the truly essential transformations, those concerning the principles, we realize that there was a long evolution accompanied by a total of three revolutions, at least in physics. To extrapolate from those episodes and formulate a rule predicting a future avalanche of revolutions, as some do, appears to be unwarranted. How many revolutions did we have? Three in all. It would be a rash conclusion to deduce, from that, "One, two, three, always." I would be careful not to predict the end of scientific revolutions, but I have the right to consider it as likely an option as its opposite.

For that, I will no doubt be labeled a conservative, and I can already hear the old line: physicists believed that science had come to an end in the late nineteenth century, precisely when its most radical transformations were in the making. Beware of ever repeating such a blunder! But who decided that next time it will also be a blunder? Would it not be wiser to avoid categorical statements and simply ask the question: Since we were wrong once, does it follow that we shall be wrong every time?

Let us remark in closing on the similarities between Thomas Kuhn's reflection and that of Michel Foucault in *Les mots et les choses*. The former considers science at a given time as an assortment of paradigms and imitations, all sharing a common source. For Foucault, it is the entire collection of intelectual achievements that is so related—what he calls the century's *épistémé*. In both cases, the rallying concepts, *épistémé* or paradigm, may be convenient indicators for a history of mentalities, but they have nothing in common with reality, the only object relevant to science.

Method

W<small>E ARE GOING</small> to pursue our examination of the state of the question, this time focusing on the method of science. The topic is unavoidable, especially since it is often denied that such a method even exists. I am thinking of course of Feyerabend and his followers. Let us take a look.

A METHOD FOR JUDGING, NOT FOR BUILDING

Given the vastness and consistency of today's science, one cannot help wondering what is the source of those attributes, and even how science itself exists at all. Reality is certainly the cause, but by which powerful method do we question it and obtain such generous and, at times, such strange answers?

Bacon or Descartes used the word "method" in its ordinary sense, a rule of behavior inexorably leading to more knowledge: a method to build science. In this sense, there is a certain contradiction between Bacon's criticism of philosophy and his belief in the power of method.

In fact, to assume that such a method exists is a philosophical postulate. A method to generate science with enough certainty presupposes the possession of a principle of a higher order than those one might eventually discover through its use. Descartes does possess such a principle: the preeminence of reason, before which everything becomes immediately clear. Bacon assumes that reality "speaks" by itself, and that it is enough to question it. This amounts to putting an almost blind trust in induction. I prefer the other alternative suggested by Bacon when he speaks of "proceeding regularly and gradually from one axiom to another, so that the most general ones are not reached till the last." This approach proposes to pluck the philosophical principles right out of the tree of experience, including the guiding principles of science.

It is not a method to build science that we shall seek, but one that can be used to judge it after it has been built, without imposing up front the form it is supposed to have. Basically, it consists in a collection of practical rules to estimate the quality of the correspondence between the scientific representation and reality; a set of criteria for testing truth, or, rather, agreement with Reality. When "method" is understood in this sense, it does not include the particular paths researchers may follow in order to gather information or make discoveries. This notion of method concerns humankind as it reflects on the accumulated knowledge more than it concerns those who seek to increase that knowledge. It is a method exclusive to science, which sets it completely apart from all other representations of reality.

WHICH METHOD?

The question of method is a highly controversial one among specialists in epistemology. The difficulty partly stems from a confusion between two related but entirely different questions: How is discovery possible? How does humankind establish the agreement between knowledge and Reality? It is the first one that leads to contention, while we are interested primarily in the second.

It is easier to begin by saying what method is not. It is not a research project, or the compilation of a database, nor a set of rules of conduct to "guide the mind" in solving problems by reducing them to a simple, even trivial, form, as Descartes believed could be done. I do not wish to give the impression that I consider such enterprises or behavior futile, either. But they result from an effort to be organized and efficient that is not particularly scientific.

What is then the scientific method, if such a thing exists? If Thomas Kuhn was right, and the advancement of science is only a succession of breakthroughs that are offered as paradigms to be imitated, I would be tempted to answer in the negative. We would then have as many methods as there are paradigms, changing with the spirit of the time and resembling inspiration more than precept. Feyerabend went even farther, and he explicitly denied the existence of method in the construction of science.

It is important that we specify what science we are talking about and what is the purpose of the method. An incipient and still inarticulate science, or an empirical one at best, cannot rule out the existence of a miraculous method that would guarantee its certainties and validate its shaky concepts. We shall therefore limit ourselves to those sciences having attained a high level of consistency, those monuments of knowledge that Roger Penrose calls "superb." They are the ones provoking amazement and inducing reflection.

As for the purpose of method, it cannot be, once again, a code of behavior accompanied by the promise to produce results: satisfaction guaranteed or money back. It is clear that having a method that would reveal the intimate nature of Reality presupposes an almost perfect knowledge of that Reality. There is no method to map out a route in unknown territory. This simple argument convinced me that Feyerabend's criticism was partly valid, if obvious. His examples confirm this impression, and so we shall abandon this once appealing but now obsolete idea.

The method we shall discuss is the one that allows us to recognize *in hindsight* whether a science is soundly established and has achieved a consistent body of knowledge.

By defining method in this way we are implicitly assuming that Reality can be known (at least in part) using criteria of universality and logical consistency. This is a very strong hypothesis, to be sure, but it nonetheless corresponds to the evidence, surprising perhaps, but irresistibly imposed by the facts and confirmed by the passing of time.

Finally, it is impossible to talk about method without mentioning Karl Popper and his principal criterion, which restricts science to the formulation of propositions that can be experimentally refuted. Popper's condition has by now become classical and may be taken for granted. The method we shall discuss fully incorporates it.

A FOUR-STAGE METHOD

There exists a well-defined method that highlights science's specificity. We shall call it the four-stage method, for it involves four different activities of experience and thought, corresponding at times, but not always, to the four stages of the history of a science. They are rather four structures of knowledge that complement each other. We shall call them empiricism, concept formation, development and verification.

This method pervades contemporary physics and is part of its "folklore," the things everyone knows but does not know where to find. Its origins may be traced back to Pierre Duhem's *La Théorie physique*, where the method is clearly presented, except for some minor details due to today's more conspicuous presence of the formal character of science. It is not very likely, though, that this book has had much influence, for it was not widely known in scientific circles; more reliable sources might be Einstein's correspondence, Heisenberg's treaties, or Richard Feynman's book *The Nature of Physical Laws*.

It turns out that the four stages in question correspond quite closely to the different periods in the history of classical mechanics that we have already discussed, making this science a convenient example. The empirical, or exploratory, stage consists in the observation of facts, the performing of experiments "to see what happens," the compiling of a catalog of data, and, eventually, the discovery of empirical rules. We recognize here the observations and measurements of Tycho Brahe and Galileo, as well as Kepler's empirical rules on planetary motions and Galileo's rules on falling bodies. It is obvious that a field of knowledge at this stage of development is not yet a mature, consistent science.

The second stage is that of concept formation or, more precisely, of conception. It consists in the development and the selection of appropriate concepts permitting a representation of Reality, the invention of the principle, or principles, that might govern this representation. We use the term "invention," and not "discovery," on purpose. Indeed, there can only be discovery after verification. As it is impossible to prescribe how to invent, this aspect of conception has never revealed its genesis. Different scientists might explain it in different ways. Some will try to justify it by a logical chain: "This example suggested that such or such concept should be a central one, and the phenomenon it represents should act in a certain way; this other example narrowed the domain of possibilities; yet another one forced me to search further. I then asked myself whether this particular hypothesis was the simplest one. . . . I've tried it and eureka!"

To this kind of logical explanation some completely irrational examples are often opposed, such as the invention by Kekulé of the cyclic structure of benzene: he saw in his dream a snake biting its tail. "But of course, the benzene molecule is a ring!" We may also say that, regardless of the particular circumstances, this is the stage of genius, in the etymological sense of the word.

We may try to imagine how Newton went through this conception phase as he formulated dynamics. First, he had to clarify the concepts of mass, force, position, and speed, and also invent that of acceleration. The latter, and not speed, was eventually to occur in a principle, but this was not at all obvious. Did he invent the law named after him after considering several possibilities, or did this law impose itself upon his mind? He said that it was the second alternative, an illumination accompanied by a sense of certainty often found in other discoverers. This kind of sudden knowledge, which we may call of the third type, as Spinoza did, is fascinating, but it may also be misleading. We shall then class it, regretfully, among the human aspects of science, a topic beyond the scope of our discussion.

The third stage, development, is on the contrary rather well guided by logic, even if occasionally the ride may be bumpy. It consists in examining all possible consequences of the principles, perhaps at the cost of a lot of effort and imagination. In most cases, only certain consequences are considered, primarily those concerning known facts. Thus, Newton begins by testing his principles on planetary motions and falling bodies. Only later, with the passing of time and the work of many people, comes the more ambitious endeavor aimed at the totality of the consequences.

In the case of physics, this development often takes the form of calculations, since the new representation to be tested is formulated in mathematical terms. This is practically never the case in biology, where common sense logic, enlightened by an extensive knowledge base, plays a role only during the reasoning process. Charles Darwin's *On the Origin of Species* is very revealing in this respect.

The fourth stage is verification. This is the stage Popper refers to when he says that a theory must be open to "falsification," that is, it must be possible to refute it. It is the phase when the theory, the idea, or the principle, until then only hypotheses armed with their predictions, are going to offer themselves to refutation. Each prediction is systematically subjected to the test of experience. What does the latter say about the prediction, is it true or false? What do the myriads (not an exaggeration in the case of quantum mechanics) of predictions and experiments say? If the answer is always yes,

259

then the idea is certainly true, and nature considers it worthy of belonging to the heart and the marrow of things. The answer is no, even a single time? It is because the theory is false, or at least incomplete. It must be rejected; at best, it cannot be trusted until new clarifications are obtained. A true verification of a "superb" theory that unifies vast domains of knowledge would be achieved only if never, strictly never, did the experiment answer in the negative within the scope of the theory. Otherwise, the edifice would waver, and could only be saved by a profound transformation that would bring about a higher level of consistency.

THE NATURE OF THE FOUR STAGES

It would be naive to expect this method to manifest itself every time a science is born or undergoes a transformation. Indeed, the method merely provides science with a frame and is not a law of history. At times, certain stages may appear to be absent, often because they are too easily completed and go unnoticed. The initial observations may be so explicit that the second stage need not require any genial insight; or the development stage may simply amount to an elementary argument. History may also render difficult the recognition of the various stages, as is often the case. Let us add that the method we have described here does not apply to mathematics; it only concerns the physical sciences, whose subject is the study of Reality under any of its forms.

Having analyzed the method, it is now easier to understand how the scientific representation is constructed and its relation to reality. Reality is summoned twice, at the beginning and at the end of the process, of which it becomes the ringleader without whose approval nothing is valid. During the exploratory phase, it provides the necessary information for the thought process to begin. Reality takes part in the verification stage, which may last several centuries, by not giving any negative answer to the succession of predictions that scientists strive to render complete. Only then can we consider our knowledge as certain, in so far as this term has any meaning.

The third stage, development, is a vast logical exercise and the one aspect that the teaching of science emphasizes the most, to the point that it might appear to the untrained eye as the archetype of the scientific method. It is partly to counterbalance that pernicious tendency that Popper insisted so strongly on verification, at the risk of destabilizing the entire building. In the construction of a representation, the main role of development is to prepare the ground for the fourth stage: verification.

The second stage, conception, is fascinating, and it enthralls all those who are more interested in humans than in the world surrounding them. Romanticism may get into the picture, marveling at the work of a genius who achieves supreme clarity as a reward for a relentless search. It is Balzac's Balthazar Claës meeting "absoluteness." This often surprising stage is also puzzling due to the occasional irruption of some illogical or irrational element (Kekulé's dream), and for the hints, the associations, and the analogies that assail the creative mind. A rationalist spectator conditioned by traditional teaching to equate science with a purely logical behavior on the part of the researcher would be stunned at the discovery of an excited mind's unpredictable course, so well portrayed by Arthur Koestler in *The Sleepwalkers*. It is highly likely, though, that these irrational aspects are the consequence of an intense intellectual activity controlling every component of the personality. Ultimately, that feverish buzzing of the mind will be swept over by silence, once the goal has been achieved. The author will then carefully tidy everything up and the result will adopt the conventional and convenient form of a scientific publication, where only the key idea remains, like Venus coming out of the sea, her feet now clean of foam.

If one has chosen to give precedence to Reality, the amazement is elsewhere. It is in the ballet of Reality with itself when the human brain, a product of Reality, generates an image of that Reality as perfect as it is unexpected.

THE LESSON OF THE FAILED ATTEMPTS

In scientific matters, history and teaching curricula usually only retain the successful attempts, thus projecting a flawed image of the discovery process. It may also appear that only exceptional individuals can "generate" science. But the most serious consequence of ignoring the unsuccessful attempts would be to give the impression that verification is a mere formality. When a human mind finds a complete explanation for some natural phenomenon or uncovers a universal principle, our sense of wonder is so absolute that we cannot conceive nature refusing to endorse such a miracle. This is far from being the case, and aborted attempts are, on the contrary, very instructive.

Particle physics is a science still in its youth, most of whose specialists are alive today. They remember the numerous theories that have been proposed in the past forty years, such or such principle whose consequences happened to be quantitatively verified by a multitude of data. Many researchers would then set out searching for new predictions that would confirm the previous findings. Experiments involving big accelerators would be set up, and often, alas, one or more of these would contradict the expected result, or produce a value of some parameter different from the predicted one. Sometimes, the effort would not be completely wasted, and the experimental results would suggest new and, eventually, successful possibilities. At worst, the regularities that the aborted theory had revealed would join the ranks of empirical rules.

It is interesting to retrace the long path that led to two major discoveries in particle physics—the unification of weak and electromagnetic interactions and the discovery of quarks—without omitting the rejected trails, appropriately described in Rilke's verses:

> Paths going nowhere
> As if diverted by chance,
> Paths that have lost their way.

The cemetery of good ideas that did not survive is immense, and not only in particle physics. I remember, not without a disdainful smile, an ironical proverb that served as an epigram to the demise of an idea that had not resisted the test of observation, an idea in cosmology I once cherished: "Nothing is more dreadful than the despicable murder of a beautiful theory by abominable facts." Let us underscore here scientists' imagination and fecundity in generating hypotheses as so many arrows shot in the direction of Reality.

METHOD AND THE SOCIAL SCIENCES

The preceding remark takes us into the domain of the social sciences. It is not our purpose to criticize them, but only to examine their methods. A great number of studies, and even entire sciences such as demography and economics, make an extensive use of mathematics, particularly of statistical methods.

It has often been said that the sciences that resort to such methods are closer to the physical sciences than those which do not. To be sure, statistics provides tools for establishing correlations, that is, the total or partial joint manifestation of two or more features. In epidemiology, to use a well-known example, a correlation has been observed between populations with a diet rich in fat and those with a high frequency of heart attacks. Correlation is often a sign of a causal relationship, but it does not explain either the cause or its action, or even if there really is a cause. In this particular example, thanks to clinical studies, to a detailed examination of the facts, and to the progress made by physiology regarding the metabolism of fats, the mere existence of a correlation can be replaced by the knowledge of the mechanisms in action, at least in part. Unlike the crude correlation, this knowledge may be subjected to the scientific method.

Basically, statistical methods are a valuable tool to accelerate the discovery of empirical rules, but it would be a mistake to assume that they are sufficient to attain the consistency afforded by the full scientific method.

The question of method has long preoccupied the specialists in social sciences, and it is particularly within this context, more than in that of the physical sciences, that Popper's analysis was developed. It is a delicate question, to which I would like to propose my modest contribution by discussing Claude Lévi-Strauss' remarkable structuralist method in anthropology. If, once again, we are led to conclude that an insurmountable hiatus exists when we compare it with the method of the physical sciences, Lévi-Strauss had already recognized it himself.

Here is a rough summary of the structuralist method, hoping I have not misrepresented it too much: one studies a certain category of facts, such as family relationships or table manners. First, in the preliminary stage, all known facts on the subject are gathered to form a so-called corpus. It is a substantial task, akin to the empirical stage of the four-stage method. The second stage, conception, is also present, for the scientist imagines, invents, or merely acknowledges (the exact word is irrelevant) a principle that organizes the facts and which is called stucture. The third stage, that of development, then consists in systematically detecting within the corpus the universal presence of the structure.

But the fourth stage, allowing the possibility for the theory to be refuted, is unfortunately missing. The reason for this absence should be clear: if the corpus is complete, no prediction can be made regarding future observations of facts, because there is nothing outside the corpus. On the other hand, it would be unreasonable to expect the researcher to totally ignore a considerable portion of the corpus that could eventually be used for verification. At best, the researcher can only hope that newly discovered facts will confirm whatever predictions were actually made. But a well-known difficulty remains: ultimately, the predictive power of social sciences is too limited.

There is another, less known difficulty: the authors' creativity may backfire and raise more doubts regarding the real existence of the structures they claim to have discovered. Anthropologists are certainly not less imaginative than the physicists who, in the 1960s, managed to find many structures in the mass of data concerning elementary particles. Yet some of these structures turned out to be illusory. What is more, the structures proposed by the physicists required quantitative verifications much more demanding than the qualitative relationships found in anthropological structures, the latter allowing more room for interpretation. Now, very little is known regarding the limits of intelligence and imagination. Isn't it possible that a sufficiently imaginative person could impose plausible structures on any given corpus? How not to be skeptical, then, when we know that the same method used elsewhere in more stringent circumstances led to illusions? The structuralist method may justify an intimate conviction, but it does not appear to allow for the irrefutable proof that only verification can provide.

CONSISTENCY AND BEAUTY

We have already said that each experiment designed to test a universal science is a throw of the dice, where the fate of a principle is at stake. This is how, almost daily, science is confronted with Reality in laboratories throughout the world. To truly appreciate this relationship between researchers and Reality, made up of admiration and provocation, it is necessary to know the joyful Machiavellianism of those who take several years to prepare a crucial experiment, aided by all the technical means they can muster or create. They belong to the "Inverse Millionaires" club of specialists who measure physical quantities to six or more significant digits, avowedly hoping to verify with even more precision the consequences of some known principle; but also, primarily, in the half-admitted hope of detecting a discrepancy that would prompt a new hunt for principles. This is how science is being built, by offering it to destruction. Its survival is the proof, repeated every day as surely as the sun rises, of the persistence of Reality and the existence of its laws.

Anything that is both profound and consistent will most certainly provoke in us a sense of beauty. This aspect of the scientific enterprise integrates one of the richest attributes of our humanity: the aesthetic one. The connection is not accidental, for we first developed our sense of beauty in the contemplation of Reality, in the enjoyment derived from the harmony of a landscape or a human face, transformed perhaps by a flute into the beauty of music, formal and tender at the same time. Beauty, when it reveals itself in a perfect balance and a supreme economy of means, is present everywhere in the great picture of science.

Even if the conceptualization of science does not obey any rules, it often becomes a quest for harmony. Dirac went as far as saying that one can first recognize a valid theory by its beauty. He was certainly referring to a form of beauty particularly valued by mathematicians, one that is hard to distinguish from consistency: "In there, all is order and beauty. . . ." Why is it that logical consistency, when it applies to so vast a domain as to produce our astonishment, creates in us the same emotion, the same kind of joy provoked by the beauty of things? I would not know, and must therefore remain silent on this mystery. It is, nevertheless, an important aspect of science, and if we try to understand the link between philosophy and aesthetics we must at least illustrate it by an example.

It concerns a principle we have mentioned several times already: the principle of inertia. How beautiful it is! At first, it appears as a modest principle regarding horizontal motion on earth. It becomes universal thanks first to Descartes and later to Newton, who ties it to absolute time and space. It frees itself from this matrix to reappear, intact, in the theory of relativity. Our craving for beauty finds its restriction to some privileged systems of reference to be an unbearable imperfection. The principle then cleanses itself by encompassing the effects of gravitation, gaining once again in generality. Like a firework display that dies out in a bouquet, the principle is today simply a law, a secondary consequence of the principles ruling the curvature of space-time: Einstein's equations of the relativistic theory of gravitation.

As Arthur Rimbaud put it: "Now that it is over I can hail beauty." These are a poet's words, expressing better than anything else the scientist's feelings.

THE FLEXIBILITY OF PRINCIPLES

Our last observation on relativity suggests another remark concerning principles. We have already mentioned that special relativity can be formulated in terms of space-time, as well as in a language that keeps time and space in separate reference frames. The first description is geometric and the second is algebraic, expressed in terms of space dilation and time contraction when going from one reference system to another. This shows that the coding of the laws of nature is not as rigid as that of the laws voted in parliament, with their definitive titles and articles. The laws of nature can be translated into other logical forms that may appear remote but are nevertheless perfectly equivalent. Quantum mechanics provides another remakable example, with its multiple versions, all equivalent, such as Heisenberg's matrices and Louis de Broglie's wave functions.

And so the form of a theory, or even its central concepts, are hardly unique; nor is there just one good question at the origin of the theory. Principles and laws, fundamental and derived concepts may be interchanged and indifferently cast for the role of a god or that of an avatar. The theory itself remains unique, for its different forms are all equivalent: they lead to the same consequences and can often be derived from each other.

It is as if the form of the principles, the path to be followed, is not imposed. There is no single trail leading to the top of the mountain. Each summit of knowledge appears as a "reality within Reality," existing by itself and accessible through each one of the faces explored by science.

THE THING IN THE WORLD MOST
EVENLY DISTRIBUTED

In closing, we shall emphasize a strong connection between the four-stage method and both the development of the formal sciences and its recent corollary, the change in status of common sense.

The contemporary scientific method would be a source of amazement for our predecessors. There is nothing like it among the methods discussed by Bacon, for whom that of induction occupies center stage. According to this ancient method, an attentive study of the facts leads almost directly to the laws they obey, thus permitting us to "induce" the laws from the facts. This is a far cry from the modern idea of a conception stage, and nothing is farther removed from Cartesian common sense than the free flight of the imagination characteristic of this stage.

We shall not insist on the relationship between formal science and common sense, already analyzed at length. Let us only note that, after Hume, contemporary psychology would recognize, as Piaget did, the origin of concepts and common sense in children (his genetic epistemology) in the observation of the world surrounding them; except that in our early years we have not watched trains traveling close to the speed of light, and our cribs were not placed near black holes where the naked eye could see space curve. We have never seen electrons diffract, either; only, at best, light moiré a spider web. No wonder then that, in the circumstances, we cannot imagine or "picture" them. Those images are missing from our repertoire, and our brain is unable to create them.

The world could have been simple and everywhere identical to what it appears to be at first sight. This is what ancient philosophers thought, and they built sweeping principles based on this belief. Science would then have remained classical, a reasonable science, and formalism would have been only a facing designed to give it more precision. Things could have been like that, but they aren't. This is a fact, and who are we to dictate to the world what it ought to be?

The wonder is elsewhere. in the power of science to tear down the walls that seemed to hold us in their grip forever. Hume was wrong in believing the human mind incapable of reaching the source of the order of things, the order that allows us to name them. Kant too was wrong in hindering our desire to understand by imposing a priori ideas and confining it to what imagination can represent with images and words. We had access by other means to that which ordinary language cannot attain or represent, that which is outside the interminably flat space, and we even had access to objects that occupy more than one place. The wonder is there. And the question then facing philosophers should be the origin of this deliverance of the mind.

The answer is clearly in the scientific method that we have just described. While the classical vision of the world is intrinsically limited, nothing restricts the scientific representation. During the conception stage, the method is free to consider all hypotheses, even the most far-fetched, in order to mimic Reality. Everything can be tried, a bold abstraction of something that has succeeded elsewhere, the exploration of the faintest clue, or a leap through empty spaces. The mountain peak where it lands has experience as its only sanction and consistency as its only ethic. This conception phase does not obey any precondition: how can we, once again, expect the world to follow our own rules? We can only set out in quest of its rules, and they are admirable. The fact that we can attain those laws through mathematics must be interpreted as a major philosophical revelation.

Thus, the method exists, boundless, its ultimate foundation being the freedom of the mind.

Vanishing Perspectives

WITH RESPECT TO what was formerly known, the new elements in the current state of the question may be summarized in three points: logic penetrates the world at the level of matter, and not at the level of our consciousness; our knowledge of the laws of reality is now sufficiently ripe for this consciousness, its intuitive and visual representation, and the common sense it harbors to appear with near-certainty as the consequence of much more general principles; finally, we are ready to accept, pending a complete inventory, that there exists an irreducible disjuncture, a chasm, between theory and reality.

That is the least a new philosophy of knowledge should take into account, together with everything else science might still supply. I believe that all is ready for the construction of such a philosophy to begin: the building blocks and other materials are there, and the plans are taking shape. We cannot afford to botch an enterprise that will undoubtedly require time and the reflections of many people. This is why at present we can only state a few hypotheses, hoping that others will soon follow, and that in challenging them, in analyzing them, and in refining them progress will ensue. I shall take the liberty of proposing some tentative directions, whose speculative character I would readily concede.

THE THEORY OF KNOWLEDGE

It is convenient to begin with the theory of knowledge, for we have already seen its broad outlines, and little needs to be added. By "theory of knowledge" I understand a scheme seeking to explain how human consciousness may know the world, a world that obeys its own laws. It is therefore a game of correlations between the world and consciousness. More precisely, the theory proposed here considers that the origin of consciousness, and of the connections it establishes with the world, lies in the laws obeyed by the latter. As for the philosophy of knowledge, it lies beyond that; it ponders the world and our awareness of it, assumed to be already understood, in order to penetrate the nature of that world. Ultimately, it could even ignore the existence of humans, who appear only as containers of thought, the flame-bearers, precious but temporary and accidental, of the universe contemplating itself.

The theory of knowledge we are led to is almost obvious. We cannot do otherwise but adopt the point of view of Hume and of contemporary cognition science, and admit that the perception of the world around us generates in our brain, in our mind, a representation sufficiently common to all of us for language to be able to communicate it, and sufficiently organized for common sense to exist. These are the fruits of an apprenticeship by each individual (by the species as a whole, too) and, even earlier, of a long evolutionary chain of other species in their adaptation to the world.

The world we apprehend is not the more fundamental one of atoms, and all the objects we perceive in it are incomparably larger. It is from this larger scale that those objects inherit the particular features whose source can be found in the universal laws that are valid on every scale—although at the atomic level these laws possess other characteristics. Thus, the world within our reach reveals itself in the form of objects that can be perceived by sight, by touch, and by hearing. We know already that the "conspicuous" world is but the subtle manifestation of the quantum laws, their metamorphosis on a larger scale. It is perhaps amusing to observe that, among our senses, smell and, to a lesser degree, taste, are detectors of molecules working on a medium scale (this is also true of sight, which may detect a very small number of photons, but only in circumstances that are too exceptional to be relevant). Be that as it may, our world also exhibits some persistent features: events can leave long-lasting traces. This is, from the point of view of physics, a form of determinism, but it is above all the possibility for memory to exist, that is, memory as traces of the past, in ourselves, as the perception of that past, and, thanks again to determinism, as the anticipation of the future. Indeed, our world often behaves in a predictable, repetitive manner, in which the laws of physics do not play a greater role than do the regularities of the living world encoded by common genetic rules. Thanks to this providential monotony, this pervasive order, we can form an interior "image" of the world and we can describe it using language.

All the features by which the world records itself on us may therefore be derived from the fundamental principles that rule the essence of reality. Such is the well-defined framework of a theory of knowledge in which those principles come first and the various forms of consciousness after. This deduction has now been essentially completed, and the theory of knowledge so obtained possesses a foundation sufficiently sound to allow the construction to proceed safely.

We can then recognize how some ancient "philosophical principles" reappear, keeping in mind that they apply only at the large-scale level. This is how intelligibility and locality, as well as discernibleness, become *properties* valid at that level. As for causality, it is closely related to our daily experience with determinism, whose limits are well known.

And so, the "philosophical problems" that the quantum laws seemed to create disappear by themselves. The domain of application of the admissible "principles" is in effect quite limited. The statement that a given principle is in action in a particular instance must always be accompanied by a certain probability of error, ridiculously small in ordinary circumstances (which is the reason thinkers in the past believed they could formulate those "principles"). As one seeks to extend them to increasingly smaller objects, this probabilty of failure increases. Once the atomic level is reached, the probability is so large that Aristotle's and Kant's principles collapse under the unbearable burden of error.

In closing, let us emphasize that the theory of knowledge we have presented here is far from complete. At best, we have laid the groundwork, and the fact that we have only discussed the sciences of matter without mentioning those of life, and have talked about the laws of particles but not elaborated on the rich complexity they generate on a larger scale, should be enough indication that the task of the cognition sciences is just beginning.

Logos

It is only now, almost at the end of the book, that we really address the philosophy of knowledge. We cannot afford to treat the subject in a hasty and necessarily inadequate manner, but only sketch some broad outlines. This incursion into philosophy must begin with a return to a question we have not fully answered: that of the nature of mathematics. This topic cannot be avoided, for the immanence of formalism demands it more than ever, and the "No one may enter here who is not a geometer" must remain carved on our pediment.

It is not necessary to repeat the arguments in favor of "mathematical realism," put forward by the supporters of a Logos existing by itself and whose nature is different from that of Reality. We have also seen nominalism and its variants, supported by an even weaker dialectic and rather limited in its justifications, except that it does not presuppose the existence of another reality.

To that I shall add only one instance in the history of contemporary science that seems revealing. Some twenty years ago, particle physics made considerable progress toward its unification, first by succeeding in the synthesis of electromagnetic and weak interactions (responsible for the beta radioactivity of nuclei, probably the heat inside volcanos, the first stage of the nuclear reactions in the sun, and the mechanisms behind the explosion of supernovae). Next came the unification of the multiple forms of strong interactions (responsible for the forces inside the atomic nucleus), forcing physicists to admit that many particles (such as the proton and the neutron) are made up of more elementary components, known as quarks. To be sure, experimental data played an essential role in those advances. But it is less well known that the theoretical efforts had been almost entirely mathematical, combining symmetry considerations (or group theory) with others arising from the geometry of abstract spaces. Nowhere else has the penetrating force of mathematics into the heart of Reality proved so prodigious, and no awl perforates so deep and so well.

The nature of the laws leaves us even more perplexed. They are extraordinarily subtle, and yet apply to objects that are, so to speak, structureless: electrons, photons, or quarks. Consider, for instance, an electron and a photon in an otherwise empty region of space. Can we imagine anything more insignificant? They are mere particles, almost nothing, monsters of simplicity compared to a grain of sand. How could each of them carry more than an elementary symbol, 1 or 0, to mark their presence or absence at that point, how could they conceal anything else? And yet, they behave according to laws whose predictions can only be obtained through long calculations on a powerful computer—and the two particles verify, with a precision of ten one-millionth, the results of those calculations. What guides those two dumb, blind balls (not balls, really: points, without a well-defined position). How do the laws act? On what do they take hold? We know absolutely nothing. Everything seems to indicate that they do not act. In Aristotle's words, they belong to the realm of power and not of action.

Once we are aware of this fact, together with the chasm we mentioned earlier and the arguments of the mathematicians regarding the absolute consistency and miraculous prolificness of their science ("It's too beautiful, it's too beautiful, but is necessary"), we are led to the conclusion that the existence of Logos is an entirely plausible hypothesis.

Thus, to the question concerning the nature of mathematics—is it part of Reality, does it exist through Reality, or does it have an independent existence? we shall answer: part of Reality? no, because of the chasm, that irreducible hiatus that separates Reality's skin from its garments; do they exist through Reality? no, because the barren poverty of the particles reduced to themselves would be unable to sustain any symbols that might conceal the laws. Hence mathematics exist by itself, as the consistency and fecundity of the fragments already discovered by the human mind suggests.

THE INSTAURATION

The profound duality we encounter here, where Logos and Reality part, as well as the very existence of that Logos, are metaphysical escapades provided by science whose consequences are not easy to estimate, too important for us to consider exploring them by embarking on some hasty reflections. We can simply raise one or two trivial questions, if only to give an idea of the immensity and the difficulties—but also the promise—of the task ahead.

Let us begin by observing a weakness of this program, one by which the very same science that suggested the enterprise may be used to call it into question. While the existence of Reality is obvious, it took at least two stages to move from the certainty of Reality to the less than certain existence of Logos. The first stage, which seems to impose itself, is the pervasive presence of laws throughout the universe. As for the second one, which we have called the chasm, that is, the ultimate irreducibility of Reality to formalism, it is more vulnerable to criticism and open to a possible evolution of knowledge. Even if some philolsophers are comfortable with such a position, their arguments rest only on principles whose fragility is all too apparent. Quantum theory is the only one that permits us to oppose Reality and Logos in pure form, so to speak; sword against sword, essence against essence. This direct confrontation is the key to their double and inexorable existence. A refutation of this argument, the loss of both its revelation and its strength, would take away our certainty. It would then be enough that some major breakthrough in physics should carry with it the disappearance of the chasm to put us back in square one.

Even if we dismiss this possibility, other difficulties might arise. The first one concerns questions of method: before science could reach the threshold of Logos it had to go through a diet of severe asceticism. It also had to partially abandon common sense, in order to discover the extent to which the unrestricted philosophical principles of the past are unreliable. Now, those are precisely the same principles used until now by the explorers of Logos. Which explorers? you might ask. Plato, of course, but especially Plotinus, the most rigorous of all in a domain where rigor is not easily exercised. I would not hesitate to add Spinoza, whose position seems to me more akin to the one presently taking shape than it might appear at first sight. We are dualists, while Spinoza is said to be a monist, but is he? Doesn't he say, in the first propositon of his *Ethics*, "By substance I understand what is in itself and is conceived through itself," a sentence in which a logician will not fail to remark the role of kneecap played by "and"? We find in it what is, a Reality, and what conceives and conceives itself, a Logos, just as we find the same dichotomy in nature, under its forms *natura naturata* and *natura naturans* (receiving and giving shape). There is certainly a lot to be learned from Spinoza, as well as from Leibniz and, obviously, more recently, from Heidegger. They all show directions to be followed, but none of them offers a reliable method to stipulate that Logos.

Another difficulty, and an eventual source of considerable puzzlement, has to do with the very notion of existence and how to grasp it. There is a problem when we attribute it to Logos, but what exactly is the problem? The idea of existence has already a fleeting quality when applied to Reality, whose various components exist during a more or less short interval of time: the things that have existed and those which will exist, do they exist? There is that which exists now, the *Dasein*, to use Heidegger's term, and there is the Being, something that the German philosopher imagines, if I am not mistaken, as an entity combining Reality and Logos. Be that as it may, to be, to exist, Logos and time, Being and Time, *Sein und Zeit*, perform a divine ballet that we humans behold without being able to pierce the secret.

One last difficulty and the most fascinating of all consists in delimiting Logos, in conceiving its scope. Science's approach, cautious, watching its every step, leads, almost against its will, to rediscovering the metaphysical extent of what was familiar ground from the beginning. But in restricting ourselves from the outset to what science may attain, to things that are verifiable and quantifiable, aren't we also restricting ourselves to knowing only that which is most dry and arid? I recall an episode from *Mahabharata*, when the hero, Arjuna, meets Shiva in the forest. At first, all Arjuna sees in front of him is a repulsive and naked ascetic; it is only after a trying ordeal that the Master of Worlds reveals himself to the young man. Similarly, Logos introduced itself to us in the ascetic nakedness of logic and mathematics, which appear to so many people as dreary, lofty, and unwelcoming. It is, nonetheless, the same name Plotinus used to speak of the Soul of the World, the object of his blissful contemplation. Would that name be misleading, used carelessly with different meanings and having Plato as the only common root? Or is it rather a clue, an opening? How far then does Logos stretch? What is its range?

As we have already said, we are merely outlining some possible paths to be explored. The book is coming to an end, and whatever solid ground we have covered is behind us. Caution is no longer required. Since we are advancing haphazardly, let us be daring.

First, we must deal with the question of method, without which nothing can be said. Unlike Reality, Logos never offers itself in a concrete form, even if it is present everywhere in the reality accessible to us. There is perhaps the beginning of an answer, a handle, so to speak, for whoever tries to get a hold of it. We may not know much about Logos, but we possess a sort of living mirror of it: the brain, which was born and evolved to accommodate it, to exploit it, to recognize it. The brain carries a trace of its matrix as a meteor carries that of an inaccessible planet. The idea is quite simple: everything our brain translates as some form of order is perhaps the reflection of a possibility of Logos. This is clearly a rough hypothesis, and it propels us imprudently toward everything we have tried so far to avoid: fuzziness, arbitrary and hastily formulated principles. A severe criticism should follow—and I confess not to know even where to start. But let us use this idea as a guide, not with the insane purpose of determining Logos' domain but to imagine perhaps the extent of that domain. That restriction, the word "perhaps," is important, because it opens up possibilities without guaranteeing them.

We have remarked several times the existence of a certain kind of beauty, cold and pure, in mathematics. Let us turn the idea around. Our brain seems capable of connecting the order and harmony it discovers in those sciences with what it perceives more generally as being beautiful. Some will say that this is only a confusion of psychological mechanisms without real significance, and having many possible causes, a few molecules of liluberin tickling the hypothalamus or some other hormonal effect of uncertain origin. But let us recall something Plotinus said about beauty. For him, the beauty of the statue of a God did not reside only in the shape of the marble proper, but also in what the artist had managed to capture of the divine nature and had offered as a reflection—the manifestation, in a concrete and real object, of a form whose natural dwelling was Logos. Using again words already employed, we could summarize this theory of beauty by saying that it is a partial representation of Logos in Reality.

At the risk of invading a domain far removed from my field of expertise, and one which only the masters of aesthetics may discuss, it appears to me that many among them have never renounced Plotinus's vision of beauty, even if they have qualified it. If, for the sake of caution, we restrict ourselves to what in aesthetics is closer to mathematics, the importance of the symmetries in a figure, the accuracy of its proportions, which are, at a more abstract level, another form of symmetry, another manifestation of what is known as groups, then all that is well known. It is also known that a discrete departure from an excess of symmetry may break the coldness of a work and introduce a kind of presence of the surrounding universe, a manifestation of life. Haven't we recently discovered in amazement that the shape of breaking waves, of clouds crossing the sky, or of a mountainous landscape could be faithfully imitated by forms resulting from calculations representing fractals—that is, mathematical objects possessing a subtle symmetry that renders them similar to themselves on whatever scale they are observed? Also, in that case, a certain lack of symmetry removes a little of the mathematical perfection to give an impression closer to reality, without destroying the impression of beauty.

Does beauty belong in some sense to Logos? Many have believed so from Plato on, and our discussion of fractals, that tongue-in-cheek message of the latest mathematics to aesthetics, renders the idea more relevant and pressing than ever. Let us try to remain scientific, tough. Beauty must be felt, it is something that manifests itself in our brain. If we succeed in retracing its source to Logos, we can also imagine that physiology can be short-circuited, that we can bypass "this brain, this greyish and fatty mass," to move right into artificial intelligence (an unfortunate expression that includes in principle the abstract structures of thought) and envisage an extension of the cognition sciences to aesthetics and, as a theoretical parallel, an exploration of the aesthetic domain of Logos. This direction, as soon as we timidly half-open it, appears immense. To borrow Bacon's words in his *Great Instauration*, one cannot expect it to be the work of a single generation.

Another indication in a similar direction was provided by Heidegger in his later works, where he proposes to find the best path toward the knowledge of Being—or, from our point of view, of Logos, at least—in the summits of poetry. Following the same approach as above we are led to some prospects that are frightening in their boldness, but not necessarily absurd: we must explore the poetic structures using the cognition sciences and artificial intelligence as a guide, in order to grasp the underlying semantic forms, symmetries, and fractures. And how about music and beyond?

Thus, whatever the extent of Logos, it may open up new domains of knowledge whose vastness seems without limit. However, it would be a mistake to see in this opportunity only the invasion of the domain of the chosen, that of art, by the narrow-minded practitioners of science, a kind of blind trespassing. On the contrary, one should see in it the consolidation, in a structured and serene form of thought, of the most beautiful dreams and the most lucid contemplations of the past. But how Cyclopean the task appears!

FOUNDING SCIENCE

I would like to finish with a question that can be very simply put: How can science exist? Or: How is science possible? The obviousness of this question and the silence surrounding it echo Aristotle's beautiful words: "Like night birds blinded by the glare of the sun, such is the behavior of the eyes of our mind when they stare at the most luminous facts."

Why is such an obvious question so seldom asked? Are scientists so indifferent to the spectacle taking place before their eyes that they only see an ordinary, trivial scene? Are they preoccupied only by their next discovery or interested only in impressing their peers? Or is it perhaps the habit of seeing each problem end up in a solution, each experiment yield a result, that leads them to a kind of unquestioned certainty, to an absolute faith? They possess in effect an unshakable faith, the stronger because never explicitly declared.

If, nevertheless, we were to ask a scientist the question, "How is science possible?", the answer, almost certainly, would be a laconic "Let us not get into metaphysics," meaning, "It is a domain of dubious reputation, and I'm not willing to risk mine—my reputation as a serious and competent scientist, that is—by being seen in such an unrespectable company." This was not Einstein's reaction, though, he who said, "It is a wonder that science should be possible." But where does this wonder come from?

The answer is perhaps as obvious as the question: science is possible because there is order in Reality. The laws that structure the representation we form of Reality are an image of its own order. The whole of science suggests such an answer, but science alone cannot establish or even formulate it, for this assertion is beyond science's own representation. Science is restricted to the region of Reality already explored; it cannot get out of it or assess it. To go beyond what is known amounts to proposing a hypothesis about the unknown, to leaving science and entering metaphysics.

For, after all, this very simple statement, "Reality is ordered," is enough to found science by turning the tables. It turns the tables because the long path we traveled in order to understand what science is now becomes clear: Reality possesses the highest possible order (but does "possible," as used here, have a sense?) or a perfect simplicity (but does "perfect" have a sense?). This order organizes

Reality from its most elementary to its most complex aspects, from the smallest to the largest scale, with one sweeping stroke. The potential for consciousness is already written in the laws that govern matter, and time might make it hatch. Science is possible because Reality's order generates the consciousness that will discover it. There is a strange resonance in Socrates' "Know thyself," which takes us to a sort of "Know thyself knowing," by which Reality knows itself in human consciousness which belongs to it.

We must emphasize that the above statement only makes sense if outside Reality there is something other than it, if there is Logos. Surely, the existence of Logos is less convincing than that of a universal order, but only the former gives a sense to the latter. The order of the world is of a logical and mathematical nature, and any kind of nominalism would be equivalent to saying that, if the universe is ordered, its order stems from an arbitrary game of hypotheses and deductions, from my own capricious choices. Or else I assume that mathematics comes from Reality (of which it forms a "superstructure"), but we have seen why this position is also untenable.

On the other hand, everything becomes clear if Logos is a consistent entity independent of Reality. The order that seemed elusive materializes in a correspondence between the two, between pure order and perpetual change. It is then natural that the representation of Reality proposed by contemporary science should pass through logic and mathematics, for they constitute the image that we reach, the representation of Logos. If our ordinary language and common sense experience this order that they express, it is because they too are the consequence. It is possible to reconsider, based on these new foundations, the old nominalism-realism controversy, that is, the ability of language to convey meaning, a problem that Russell still recently considered to be relevant and urgent.

The separation of Logos and Reality thus appears both as the most appealing hypothesis and the one promising to be the most fruitful. It is also the only one that seems in agreement with what we have called the chasm, the ultimate gap between reality and its theoretical description. It allows us to conceive their correspondence as a partial penetration of one into the other that gives instant meaning to the key sentence "Reality is ordered," which in turn answers the necessary question: how is science possible? The image of this penetration, which manifests itself at the level of our

Figure 3. Representations of Reality and Logos.

representations, is simply the role of mathematics in the constitution of science. Its outrageously formal aspects are then no longer surprising.

All that constitutes a metaphysical scheme from which a new philosophy of knowledge might be built. Its structure is best summarized by a diagram than by long explanations (fig. 3).

In this diagram, science is a representation of Reality; mathematics and logic are representations of Logos. Each representation progresses thanks to the efforts of humans; it inherits from humans its fringe of uncertainty, its advances, its hesitations. In spite of that we can witness the increasing role of mathematics and logic (which are representations) in structuring science (itself a representation). This may be interpreted as the reflection—the representation, in fact—of a higher, intrinsic correspondence between the primal entities, symbolized by the line joining them in figure 3.

It would be tempting to try to elaborate on the nature of that connection, but I hesitate to do so. Everything I could say seems to me hopelessly inadequate, dubious, or, more seriously and more philosophically put, premature.

All parallel lines in a painting appear in perspective to converge toward a common vanishing point, and the perspectives I present here are no exception. The horizon tends to become indistinct near that point, considered by geometers to be at infinity and to which I have undoubtedly come too close. I had better stop here, and leave to others the task of pursuing, improving, correcting, or tracing other paths. Youth can be trusted with all that, so that it may listen, through the rough tunes that it was occasionally necessary to play, the continuo of a song of hope. For it is irrelevant whether the handful of ideas proposed in this last chapter are interesting or not. What matters is to know that we are moving ahead, that there will be celebrations of the mind, and that, perhaps, philosophy may soon start again.

✧ *Glossary* ✧

THROUGHOUT this book we have tried to avoid the use of technical or scholarly terms, which often serve only to muddle the message if the recipient is not familiar with them. A certain number of specific terms have nonetheless found their way into the text (we have indicated their first occurrence with an asterisk). Their complete list appears below, with each term followed by a brief definition. An asterisk within a definition refers to another term on the list.

Axiom — Originally, it was a mathematical proposition whose truth was self-evident. In contemporary usage, it is a proposition belonging to a *formal language** that is assumed to be true by hypothesis.

Cartesian project — In philosophy, this is the name given by Heidegger and Husserl to theoretical physics' founding hypothesis stretched to the limit, assuming that physical reality can be completely described using mathematical rules.

Chasm — This is a term introduced in this book to designate the impossibility for a theory to describe all aspects of physical reality. The gap between theory and reality stems from a conflict between the uniqueness of facts and the essentially probabilistic character of quantum theory. It refers to perfectly visible facts, and not, as in the veiled Reality proposed by d'Espagnat, to properties that are only conceivable and cannot be assigned a truth value.

Commutativity — In mathematics and in quantum mechanics, to obtain the product AB of two *operators** A and B, one must first apply the operator B to a given function u to form the new function Bu, and then apply the operator A to the latter, which produces ABu. This defines the action of AB on u. The operators A and B commute when $AB = BA$. In general, the difference of the two products $AB - BA$ is called the commutator of A and B.

Decoherence — In quantum mechanics, decoherence is a physical effect due to which quantum *interference** effects between states that are distinct at the macroscopic level disappear very quickly.

Denkbereich — See *Domain of propositions*.

Diffraction — In optics, diffraction phenomena manifest themselves by corrections to the linear propagation of light which show its undulatory nature. Thus, the edge of the shadow produced by a source reduced to a point is not completely sharp when closely observed.

Domain of propositions — In logic, it is the totality of the propositions under consideration for the purpose of reasoning in a given context. It may be defined using sets as Boole did, or constructed by means of a more or less formal language.

Empirical rule — This is a (possibly quantitative) rule, derived only by empirical observation within a class of phenomena, whose explanation in terms of *laws** is not known.

Energy — In classical physics, energy is a physical quantity that remains constant in any isolated system. It often has two components, one depending only on velocity (kinetic energy) and another on position (potential energy). In quantum mechanics, energy is an *observable** also called the hamiltonian.

Ether — This is a hypothetical medium assumed to fill all of space, whose existence was postulated by classical physics. Originally, it was supposed to provide a medium for light to propagate. Later, when light was identified with a vibrating electromagnetic field, ether also provided a medium for this field to propagate. It disappeared as a scientific concept following Michelson's experiment.

Formal — The attribute formal, as used in this book, denotes the opposite of intuitive, representable, visual, or expressible by words in the language of common sense. More precisely, a concept about reality (in physics, for instance) is considered as formal if it is expressible or can be grasped only through mathematics. Logic and mathematics are formal on the first level when they deal only with relations, and not with meaningful objects that are completely and uniquely defined (for example, a proposition about a relation among straight lines (meaningful objects) is strictly equivalent, according to the theory of polarities, to a proposition about points, as meaningful objects). Mathematics and logic may be considered as purely formal when their foundation is completely reduced to an *axiom system** in some *formal language.**

Formal language — In logic and mathematics, a formal language consists in a set of symbols and another set of precise rules specifying how the symbols may be combined to form propositions. The latter are not assumed to refer to reality or to have a unique meaning.

History — In quantum mechanics, a history is a sequence of various properties taking place at successive instants of time.

Interference — In optics and quantum mechanics, when a wave may follow two different paths (through one or the other of two slits, as in Young's experiment, for instance) its intensity (or, in the quantum case, its probability of occurrence) varies from one place to another and shows maximum and minimun values (bright and dark fringes, in the case of light), the existence of which constitute the interference phenomenon. Basically, it is due to a superposition principle according to which the amplitudes of waves that have followed different paths are added together.

Interpretation — In physics, interpretation, as we define it in this book, is the process of deriving, from the formal principles of a theory (relativity or quantum mechanics) a logical representation of observable reality in a form that is compatible with common sense and which may be communicated in ordinary language; it must also conveniently describe the experiments that are performed in practice.

Law — In science, a logical consequence of the *principles** that is confirmed by experience.

Maxwell's equations — In physics (electrodynamics) it is a set of equations that govern the properties of the electric and magnetic fields and their evolution in the course of time.

Metalanguage — A metalanguage is a formal language that gives a larger meaning to another formal language. The propositions of the latter then become words (signs) of the metalanguage.

Modus ponens — In logic, the possibility of beginning a new proof with a *theorem** that has already been proved without having to justify the proof of the latter.

Momentum — In classical physics, momentum is the product of mass times velocity. In quantum mechanics, each of the components of the momentum vector is an *observable**, that is, an *operator** involving the differentiation operation. Therefore, in this case, it is a very formal notion.

Objectivity — A phenomenon, a concept, or a piece of knowledge is declared objective, to different degrees, if its existence does not depend on the human mind. This notion was introduced by Kant; it has been studied by the social sciences and only began to pose a problem in physics with the advent of quantum mechanics. Then a question arose regarding the objectivity of certain concepts, in particular that of the *wave function.** Are they directly associated with physical reality, or do they exist only through our awareness of them? Bohr first, and modern researchers later, pronounced themselves in favor of the objectivity of the theory.

Observable — In classical physics, the basic physical quantities are the coordinates of position and momentum. Then, a general physical quantity such as energy is a function of those coordinates. In quantum mechanics, the role of a physical quantity is played by an *operator** possessing certain mathematical properties (such as hermiticity), which is called an observable. This is one of the theory's most formal aspects.

Operator — In mathematics and in quantum mechanics, an operator A is a mathematical operation which, acting on a given function u (usually a wave function) generates another function, denoted Au. Linear operators, by far the most important ones, are those that preserve the sum of two functions and the product of a function by a constant.

Paradigm — In epistemology, this is a notion introduced by Thomas Kuhn. A paradigm is a remarkable scientific breakthrough that becomes a model to be imitated by other researchers. To the explanation of the evolution of research in terms of paradigms has been opposed that of the advancement of science in terms of principles. The word "paradigm," originally not very well defined, may be found nowadays in a multitude of jargons.

Positivism — In philosophy, this is a doctrine proposed by Auguste Comte and followed by his emulator John Stuart Mill. In epistemology, it designates the point of view according to which the criterion for true knowledge is a consensus among the people concerned (assumed to be acting in good faith, to have the required qualifications, etc., with all the difficulties that the verification of such conditions entails). In quantum mechanics, it is principally the doctrine denying the objective reality of the wave function and claiming that this function represents only the information available to the observer.

Pragmatism — In its strong sense, it is Hume's philosophical doctrine, according to which facts come first and are at the origin of thought and language, the source of the order governing them being in principle inaccessible.

Principle — In science, this is a universal proposition controlling physical reality.

Principle of complementarity — This is a principle in quantum mechanics formulated by Bohr. According to it, in describing physical reality, certain incompatible notions cannot be employed at the same time—for instance, position and velocity for a particle, or the field and the corpuscular nature of light. In recent versions of the theory, this restriction remains, but only as a consequence of other principles.

Principle of inertia — This is one of classical mechanics' fundamental principles. In the form given by Newton, it states that the center of mass (also known as the center of gravity) of a body which is not subjected to any force moves in absolute space along a straight line, with uniform velocity with respect to absolute time. The same property is valid in every (Galilean) reference system, itself moving with uniform velocity and without rotation with respect to absolute space. In the special theory of relativity, the principle of inertia applies in Galilean reference systems moving without rotation and with uniform velocity with respect to each other. These form a class independent of absolute space and time.

Principle of minimal action — In physics, this is a principle from which the equations of motion of a classical system can be deduced. Introduced by Lagrange in the eighteenth century and extended by Hamilton, it states (in its simplest case) that motion minimizes the value of a certain integral, known as action, which may be calculated from a knowledge of both kinetic and potential energies.

Projector — In mathematics, and particularly in its applications to quantum mechanics, a projector (P) is a particular kind of operator. When it acts on a function u (a wave function, say), it generates another function v, denoted Pu. The main characteristic of P is to remain the same upon iteration: $P^2u = Pu$. This property is also possessed by the projection of a point in three-dimensional space on a plane, hence the name projector. The quantum observable (the physical quantity) associated to P can only take on the values 1 or 0, similar to "true" and "false." From this fact results the important role played by these operators in questions involving logic.

Property — In quantum mechanics, a property means that a certain physical quantity (an *observable**) falls within an interval of possible values at a given instant. Properties are the basic elements of any description of physics.

Propositional calculus — In logic, it is the manipulation of propositions of a certain *formal language** with the help of logical operations such as "not," "and," "or," and the introduction, among these propositions, of equivalence or implication relations.

Realism — The various forms of realism are doctrines belonging to the philosophy of knowledge. *Platonic realism* assumes the existence of a world of Ideas more real than our own world. A similar position, *mathematical realism*, believes in the independent existence of an entity that mathematics explores but does not create. *Physical realism* comes in many different forms. They all postulate the existence of a physical reality independent of the human mind (as opposed to idealism) and, often, also admit that this reality may be known in itself (contrary to what both positivism and representationism maintain). The difficulties of reconciling realism and quantum mechanics prompted Bernard d'Espagnat to introduce the idea of "veiled reality," which restricts what aspects of reality may be known.

Schrödinger's equation — In quantum mechanics, Schrödinger's equation expresses the variation of the *wave function** as a function of time, and in this sense it plays the role of dynamics. It incorporates in an essential way a particular *observable,** the hamiltonian, or *energy.**

Scientific revolution — This is a notion, introduced in the history of science by Thomas Kuhn, that designates the discontinuous changes that take place following major scientific discoveries. Each of these is associated by Kuhn with the emergence of a new *paradigm,** marking the break with the past provoked by the given "revolution." From the perspective of the *principles** of science, such a "revolution" often consists in a revision and an extension of these, within some specific domain of application. The former principles, reappearing as consequences of new ones, then acquire the status of *laws*.*

Space-time — In physics, this term denotes the conjunction of space and time in one single system, conceived as a primal entity and represented by an abstract, four-dimensional mathematical space. There are many ways to introduce coordinates in this abstract space, each of which imposes a particular structure to space and time that can be empirically verified by an observer in his or her own vicinity.

Spin — This is a quantity that characterizes a quantum system, akin to an angular momentum. It is a vector of which only its magnitude and one component can be specified, both values being multiples of $h/4\pi$, where h is Planck's constant. In a macroscopic system, the spin indicates whether or not the system is rotating on itself, but this interpretation is not valid for a particle.

Theorem — In logic and mathematics, a theorem is a proposition whose truth has been established by a proof or demonstration under the assumption that all the *axioms** are true.

Truth — In logic, what characterizes truth is the possibility of assigning a value 1 (true) or 0 (false) to a proposition. In logic and mathematics, the *axioms** are assumed to be true by hypothesis, and the *theorems** are propositions whose truth is established a fortiori by means of a proof or demonstration. In the physical sciences, and especially in physics, observed facts are considered to be true. In quantum mechanics, there are *properties** that are true without being directly observed facts, but which are a consequence of those facts.

Uncertainty relations — Discovered by Heisenberg, these relations do not constitute, as it is said sometimes, a principle of quantum mechanics, but are a consequence of those principles. The best known case involves the statistical uncertainty Δx of a position coordinate x and the uncertainty Δp of the corresponding momentum component: the product $\Delta x \, \Delta p$ can never be less than $h/4\pi$, where h is Planck's constant. As a consequence, wave functions leading to increasingly precise values of x produce at the same time values of momentum that are increasingly uncertain.

Universe of discourse — See *Domain of propositions*.

Wave function — In quantum mechanics, the state of a system is defined as datum, or information, from which the probability of every *property** may be calculated. This information is often expressed in mathematical form by a function (the wave function) whose arguments are the coordinates of the particles making up the system. Thus, the wave function is a formal quantity containing what is needed to express everything that can be said of the system at any given instant.

Wave function reduction — This is one of the main hypotheses for the interpretation of quantum mechanics according to Bohr. After measuring some quantum physical system (an atom, for example) using a measuring device, the system's wave function is supposed to change suddenly. Its new expression is then determined by the result of the measurement as indicated by the instrument. The reduction, viewed simply as a practical rule for the calculation of probabilities, remains valid in the most recent versions of the interpretation, without having to be considered as the consequence of any particular physical effect.

❖ Name Index ❖

⊹ Subject Index ⊹

absolute space, 32
action: at a distance, 35; as a mathematical quantity, 35
algebra, 50–52
antinomies: in Kant's philosophy, 73, 74, 75n
aporia, 17
astronomy, 23–28
atomic spectrum, 139
axioms, 49, 50, 96, 283; of arithmetic, 98–99

Brownian motion, 136

Cartesian project, xxii, 66, 214, 283
chasm, xxii, 211–14, 283
classical science, 83
cognition sciences, 67–69, 72, 220
collective variables, 185
common sense, 164, 165–68
commutativity, 175, 283
complementarity, 153–55, 181, 182, 222, 286
conjecture, 119
consistency, of axiom systems, 96
correspondence, 187, 188
curvature, 133, 134

datum, of a measurement, 208–9
decoherence, 199–208, 283
Denkbereich. See universe of discourse
determinism, 168–70, 190–93
diffraction, 38, 39
displacement current, 44
domain of propositions, 283. See also universe of discourse

eccentric, 25, 218
electromagnetic field, 42
electromagnetism, 39–46
electron, 135
empiricism, 66
energy, 284; of mass, 130

epicycle, 25, 218
epicycloid, 25
épistémé, 254
ether, 38, 44, 126–27, 284

fact, 223
formal, 284; approach to science, 82, 83; logic, 86–88; mathematics, 85
formalism (in the philosophy of mathematics), 117
forms, Platonic, 10, 11
fracture, 81, 82
frequency, and probability, 155
function: continuous, 59; logical, 15; wave, 142, 288. See also wave function reduction

geodesic, 133
geometry: analytic, 51; modern, 57; non-Euclidean, 53, 54, 132, 133
gravitation: in Newton's theory, 33–35; in the theory of relativity, 130–34

histories, 177–79, 284; consistent, 179, 230–33

ideas: a priori, 72; Platonic, 10–12
implication, 15, 88, 165, 180
induction: electromagnetic, 41–43; method of, 39, 40
inference. See implication
infinite series, 58
infinity, 58–61, 97–100
integral, 54
integral calculus, 52
interference, 38, 179, 196–98, 284
interpretation, 149, 150, 285
intuitionism, 117, 118

kinematics, 165

law, 70, 248–50, 285
logic, 6–22, 86–88; in quantum mechanics, 180–83

295